Risk Management: Ecological Risk-Based Decision-Making

Risk Management: Ecological Risk-Based Decision-Making

Edited by

Ralph G. Stahl, Jr.
DuPont Company

Robert A. Bachman
Maryland Department of Natural Resources (retired)

Anne L. Barton
USEPA (retired)

James R. Clark
ExxonMobil Research and Engineering

Peter L. deFur
Virgina Commonwealth University

Stephen J. Ells
USEPA

Charles A. Pittinger
The Proctor & Gamble Co.

Michael W. Slimak
USEPA

Randall S. Wentsel
USEPA

SETAC Books

Current Coordinating Editor of SETAC Books
Andrew Green
International Lead Zinc Research Organization
Deptartment of Environment and Health
Durham, NC, USA

Publication sponsored by the Society of Environmental Toxicology and Chemistry (SETAC)

Cover design by Michael Kenney Graphic Design and Advertising
Indexing by IRIS

Library of Congress Cataloging-in-Publication Data

Risk management: ecological risk-based decision-making / edited by Ralph G. Stahl Jr...{et al.}.
      p.   cm. -- (SETAC technical publications series)
      Includes bibliographic references and index.
      ISBN 1-880611-26-0
         1. Environmental risk assessment. 2.Environmental management. I. Stahl, Ralph G.,
1953–

      GE145 .E36 2001
      363.7'02--dc21

                            00-020742

© 2001 Society of Environmental Toxicology and Chemistry (SETAC)
This publication was printed on recycled paper using soy ink.
SETAC Press is an imprint of the Society of Environmental Toxicology and Chemistry.
No claim is made to original U.S. Government works.

International Standard Book Number 1-880611-26-0
Printed in the United States of America
06 05 04 03 02 01 00 99   10 9 8 7 6 5 4 3 2 1

The paper used in this publication meets the minimum requirements of the American National Standard for Information Sciences—Permanence of Paper for Printed Library Materials, ANSI Z39.48-1984.

Reference Listing: Stahl Jr RG, Bachman RA, Barton AL, Clark JR, deFur PL, Ells SJ, Pittinger CA, Slimak MW, Wentsel RS. Risk Management: Ecological Risk-Based Decision-Making. 2001. Pensacola, FL: Society of Environmental Toxicology and Chemistry (SETAC). 222 p.

SETAC Publications

The publication of SETAC Books was established by the Society of Environmental Toxicology and Chemistry (SETAC) to provide in-depth reviews and critical appraisals on scientific subjects relevant to understanding the impacts of chemicals and technology on the environment. The books explore topics reviewed and recommended by the Publications Advisory Council and approved by the SETAC Board of Directors for their importance, timeliness, and contribution to multi-disciplinary approaches to solving environmental problems. The diversity and breadth of subjects covered in the series reflect the wide range of disciplines encompassed by environmental toxicology, environmental chemistry, and hazard and risk assessment. These volumes attempt to present the reader with authoritative coverage of the literature, as well as paradigms, methodologies, and controversies; research needs; and new developments specific to the featured topics. The books are generally peer reviewed for SETAC by acknowledged experts.

SETAC publications include Technical Issue Papers (TIPs), workshop summaries, workshop proceedings, Issue Papers, newsletter (*SETAC GLOBE*), and journal (*Environmental Toxicology and Chemistry*) articles. These pubications are useful to environmental scientists in research, research management, chemical manufacturing and regulation, risk assessment, and education, as well as to students considering or preparing for careers in these areas. The publications provide information for keeping abreast of recent developments in familiar subject areas and for rapid introduction to principles and approaches in new subject areas.

SETAC would like to recognize the past SETAC Special Publication Series editors:

C.G. Ingersoll, Midwest Science Center

U.S. Geological Survey, Columbia, MO

T.W. LaPoint, Institute of Applied Sciences

University of North Texas, Denton, TX

B.T. Walton, U.S. Environmental Protection Agency

Research Triangle Park, NC

C.H. Ward, Department of Environmental Sciences and Engineering

Rice University, Houston, TX

Contents

List of Figures

List of Tables

Acknowledgments

Although we acknowledge the contributions of a number of individuals, groups, and associations in helping to compile and publish this book, we also need to thank the many people who contributed ideas, thoughts, and comments on the issue of ecological risk management and not necessarily on the contents of this book. Too numerous to name, we nonetheless owe them more than we can express in a few sentences. We are also indebted to the risk managers in various organizations who served as our unofficial sounding and "validating" boards; they were as much of the inspiration for this book as any group or individual we worked with over the last several years. A special thanks to all of the dedicated individuals in the SETAC North America Office who contributed a great deal of their time and effort to see that this book was published.

Finally, I'd like to take the opportunity to thank all of my co-editors, chapter authors and participants at our Williamsburg, VA workshop. All of them did an outstanding job of moving the ideas, and the book, along even when other, more pressing (or fun) duties were upon them. In addition, we owe a word of thanks to Drs. Jack Gentile (University of Miami) and Peter Caulkins (USEPA) for their thorough review of the book and valuable comments.

The Workshop on A Framework for Ecological Risk Management was made possible through the financial support of the following organizations:

     American Health Council

     American Petroleum Institute

     Chemical Manufacturers Association

     Exxon Company

     ILSI Risk Science Institute

     The Proctor & Gamble Company

     SETAC/SETAC Foundation

     U.S. Environmental Protection Agency

-Ralph G. Stahl, Jr.

Ralph G. Stahl Jr., a native of Houston, Texas, received his PhD from the University of Texas in 1982 and is currently a Senior Consulting Associate in the DuPont Company. At DuPont he is responsible for leading ecological risk and natural resource damage assessments. He is a Diplomate of the American Board of Toxicology and is active in SETAC coordinating the development of the Society's Technical Issue Papers and serving on the Technical Committee. He and his family currently reside in Wilmington, Delaware, where in his spare time he enjoys tying flys, fly fishing, and watching his son play soccer. Ralph is an editor of a 2000 publication of SETAC, *Natural Remediation of Environmental Contaminants: Its Role in Ecological Risk Assessment and Risk Management.*

Robert A. Bachman received a BS from the U.S. Naval Academy in 1956 and a BS from the University of Washington in 1963. He received his PhD in Behavioral Ecology from Pennsylvania State University in 1982. He served in the U.S. Navy from 1956 to 1975 including service on five submarines: two diesel and three nuclear powered. He served as exchange officer from 1970 through 1974 with the Canadian Armed Forces in Halifax, Nova Scotia and submarine advisor to the NATO Commander, Canadian Atlantic Command in Halifax. He has had 13 years of service with Maryland state government in various capacities including Director of Fish and Wildlife and Director of Fisheries Service, Maryland Department of Natural Resources. In these capacities, he functioned as administrator and risk manager for state public trust resources. He retired on 1 April 1999.

Anne L. Barton joined the USEPA in 1975 as a biostatistician and worked mainly in cancer risk assessment for five years. As director of the Environmental Fate and Effects Division in the Office of Pesticide Programs she was responsible for the ecological risk assessments (ERAs) for pesticides and developed an interest in between ERA and risk management. She co-founded a group to try to improve communication between ecological risk assessors and USEPA decision-makers, led an USEPA workgroup to produce Priorities for Ecological Protection: An Initial List and Discussion Document for USEPA that provides guidance for risk managers in ERA, joined the Multi-Stakeholder Ecological Risk Management Dialogue Group that later became the SETAC group whose work led to this book, and co-chaired a workgroup to develop guidelines for risk managers during the planning stage of ERA. She has continued to contribute to both the SETAC and the USEPA projects since her retirement from USEPA.

James R. Clark, a Distinguisthed Scientific Associate with Exxon since 1992 after a 12-year career as a research biologist with the USEPA, has a BS in Fisheries from the University of Michigan and a MS and PhD in Zoology and Aquatic Ecology from Virginia Tech. He has experience in laboratory and field assessments of petroleum industry products and activities, complex effluents, contaminated soils, sediments, and pesticides and industrial chemicals. He developed/applied ecological hazard and risk assessment approaches to address environmental issues. He was responsible for assessments of bioremediation technology developed/applied during the Alaskan Oil Spill Cleanup. Currently, he leads in the development and evaluation of environmentally relevant techniques and strategies for oil and chemical spill response. He is active in several professional and technical organizations and was elected to the SETAC Board of Directors to serve a 3-year term (1999-2001). Dr. Clark has authored over 66 peer-reviewed publications and 85 presentations at national meetings and symposia.

Peter L. deFur is an independent consultant and an Affiliate Associate Professor in the Center for Environmental Studies at Virginia Commonwealth University. Dr. deFur received BS and MA degrees in Biology from the College of William and Mary, Virginia and PhD in Biology (1980) from the University of Calgary, Alberta where he held a postdoctoral fellow in neurophysiology in the Department of Medicine. He conducts research on environmental health, water quality, and environmental policy. He is on the Board of the Science and Environmental Health Network (SEHN) and President of the Association for Science in the Public Interest. He previously held faculty positions at George Mason University and Southeastern Louisiana University and was a senior scientist with the Environmental Defense Fund (EDF) in Washington, DC. He is an editor of a 1999 SETAC Publication, *Endocrine Disruption in Invertebrates: Endocrinology, Testing, and Assessment.*

Stephen J. Ells is a Senior Environmental Scientist with the USEPA's Superfund Program in Arlington, VA. He received his BS in Biology from Villanova University (1972) and his MS in Fisheries Biology from the University of Connecticut (1974). Steve has been at the USEPA for over 20 years, starting as a Biologist in the Toxics program and then spending the last 12 years in Superfund where he was a Section Chief for over five years. Steve is an expert in ecological risk assessment and in remedy selection. He is one of the co-chairs of the multi-stakeholder group leading the development of ecological soil screening levels and is currently the team leader of the Superfund Sediments Team. He champions the use of risk-based decision-making and was lead author of Superfund's 1999 Guidance: Ecological Risk Assessment and Risk Management Principles for Superfund Sites. He is an editor on a 2000 SETAC publication, *Natural Remediation of Environmental Contaminants: Its Role in Ecological Risk Assessment and Risk Mangement.*

Charles A. Pittinger earned his BS in Biology from Notre Dame (1975), MS in Ecology from the University of Tennessee (1978), and PhD in Zoology from Virginia Tech (1984). He is currently at Procter and Gamble Co. as a principle scientist where he focuses on risk assessment and risk management development and application. He is on the USEPA Science Advisory Board, the American Industrial Health Council's Ecological Risk Assessment Committee, and SETAC's International Program Committee, Peer Review and Technical subcommittees, and Ecological Risk Assessment Advisory group, and is Contributing Editor to EcoRisk Assessment in the SETAC Globe. In 1993-1994, he was the SETAC Congressional Science Fellow assigned to the U.S. House of Representatives Committe on Science, Space, and Technology focusing on risk legislation.

Michael W. Slimak is the Associate Director for Ecology in the USEPA's National Center for Environmental Assessment where he is responsible for developing and implementing a research program to assess the ecological risks associated with multiple stressors such as chemicals, habitat loss, loss of biodiversity, and global climate change. With over 25 years of experience at the USEPA, Michael is a recognized authority on ecological risk assessments, has authored numerous government-sponsored reports, and has published in peer-reviewed journals and books. He holds a BS in Biology (1969) from Southern Nazarene University and an MS in Wildlife Ecology (1975) from Oklahoma State University. He is currently a PhD candidate at George Mason University, Fairfax, VA.

Randall S. Wentsel is the Waste Research Manager in the Office of Science Policy, Office of Research and Development (ORD) at the USEPA. In this position, he leads planning and coordination efforts for the ORD waste research program. His research experience is in ecological risk assessment and environmental toxicology. He was the first SETAC Science Fellow (along with Pittinger) and served on the US Senate Environment and Public Works Committee. He is on the Board of Directors of SETAC and is active in the Society of Risk Analysis. Randy received his PhD in Environmental Toxicology from Purdue University in 1977.

Other SETAC books these editors have contributed to

Ecotoxicology and Risk Assessment for Wetlands
Lewis, Mayer, Powell, Nelson, Klaine, Henry, Dickson, editors,1999

Endocrine Disruption in Invertebrates: Endocrinology, Testing and Assessment
deFur, Crane, Ingersoll, Tattersfield, editors, 1999

Multiple Stressors in Ecological Risk and Impact Assessment
Foran and Ferenc, editors, 1999

Reproductive and Developmental Effects of Contaminants in Oviparous Vertebrates
DiGiulio and Tillitt, editors, 1999

Ecological Risk Assessment Decision-Support System: A Conceptual Design
Reinert, Bartell, Biddinger, editors, 1998

Ecotoxicology Risk Assessment of the Chlorinated Organic Chemicals
Carey, Cook, Giesy, Hodson, Muir, Owens, Solomon, editors, 1998

Principles and Processes for Evaluating Endocrine Disruption in Wildlife
Kendall, Dickerson, Geisy, Suk, editors, 1998

Ecological Risk Assessment for Contaminated Sediments
Ingersoll, Dillon, Biddinger, editors, 1997

Public Policy Application of Life-Cycle Assessment
Allen and Consoli, editors, 1997

A Conceptual Framework for Life-Cycle Impact Assessment
Fava, Consoli, Denison, Dickson, Mohin, Vigon, editors, 1993

A Technical Framework for Life-Cycle Assessment
Fava, Denison, Jones, Curran, Vigon, Selke, Barnum, editors, 1991

List of Attendees and Participants*

Charles M. Auer
USEPA
Washington, DC

Robert A. Bachman[1]
Maryland Dept of Natural Resources
Annapolis, MD

Anne L. Barton[1]
USEPA Office of Pesticide Programs
Washington, DC

Richard B. Belzer
Office of Management & Budget,
Office of Information & Regulatory
Affairs
Washington, DC

Mike Blum
Washington State-Dept of Ecology
Toxics Cleanup Program
Olympia, WA

Peter P. Brussock
Environmental Liability Management,
Inc.
Doylestown, PA

Walt Buchholtz
Exxon Chemical Company
Houston, TX

Larry F. Champagne
TX Natural Resource Conservation
Commission-Pollution Cleanup Division
Austin, TX

David P. Clarke
The Risk Policy Report
Bethesda, MD

Mark Davis
Coalition to Restore Coastal Louisiana
Baton Rouge, LA

Peter L. deFur
Virginia Commonwealth University
Richmond, VA

James Donald
USEPA-Office of Environmental Health
Hazard Assessment
Sacramento, CA

Elaine J. Dorward-King
Kennecott Utah Copper Corporation
Magna, UT

Thomas C. Eagle
NOAA-National Marine Fisheries Service
Silver Spring, MD

Stephen J. Ells[1]
USEPA Superfund Program
Washington, DC

Susan A. Ferenc
USDA-Office of Risk Assessment &
Cost-Benefit Analysis
Washington, DC

Drew Fillingame
USEPA Gulf of Mexico Program
Stennis Space Centre, MS

Nancy Finley
USFWS
Edison, NJ

Jessica Glicken
ecological planning and toxicology, inc.
Albuquerque, NM

Robert L. Graney
Bayer Corp Environmental Research
Stilwell, KS

Eva Hoffman
USEPA-Region 8
Denver, CO

---

*Affiliations were current at the time of the workshop
[1]Steering Committee Member
[2]Breakout Group Chair

U. Gale Hutton
USEPA-Region 7
Water, Wetlands, Pesticides Division
Kansas City, KS

Patricia King
King Research & Law
Madison, WI

Mark L. Kraus
National Audubon Society,
Everglades System Restoration
Miami, FL

Robert T. Lackey
USEPA-Western Ecology Division
Corvallis, OR

Susan Lewis
USEPA-Chief Insecticide Branch
Registration Division
Washington, DC

Eugene R. Mancini
ARCO Co- Environmental Protection
Division
Los Angeles, CA

Lynn S. McCarty
LS McCarty Scientific Research &
Consulting
Oakville, Ontario

Janis McFarland,
Novartis Agricultural Stewardship
Program
Greensboro, NC

Paul M. Mehrle
ENTRIX, Inc.
Houston, TX

Reo Menning[2]
American Industrial Health Council
Washington, DC

Gary Myers
Tennessee Wildlife Resource Agency
Nashville, TN

D. Warner North
Northworks, Inc.
Delmont, CA

Charles A. Pittinger[1]
The Procter & Gamble Co.
Cincinnati, OH

Lee Salamone[2]
The Chemical Manufacturers Assoc.
Arlington, VA

Deanna Sampson
Virginia Conservation Network
Richmond, VA

Greg Schiefer
SETAC/SETAC Foundation
Pensacola, FL

Michael W. Slimak[1]
USEPA National Center for
Environmental Assessment
Washington, DC

Ralph G. Stahl, Jr.[1]
E.l. DuPont de Nemours & Co.
Wilmington, DE

Randall S. Wentsel[1]
U.S. Army
APG, MD

James E. Wilen
University of California at Davis
Davis, CA

*Affiliations were current at the time of the workshop
[1]Steering Committee Member
[2]Breakout Group Chair

Preface

As the world enters a new millenium, the U.S. and other developed countries are in the midst of substantial discussion on the management of natural resources, urban sprawl, global warming, ecological risks, and other pressing environmental issues. While the concern over the unfettered release of chemical contaminants from industrial and municipal activities in the U.S. has begun to subside since it arose in the late 1950s, local, regional, and state governments, as well as environmental and conservation organizations find a substantial challenge today in attempting to preserve valued habitats and greenspace, to provide for recreational opportunities for an ever-expanding populace, and to do all of this with a high degree of financial stewardship.

In the private sector, businesses now recognize that good environmental policy makes good business sense, and many have worked effectively in the last 10 to 15 years to reduce their environmental footprint as well as place environmental performance goals into a host of their key business decision-making activities. A simple comparison of a company's 1970 annual report to stockholders with that of 2001 illustrates how important environmental performance has become to small and large companies alike. For example, the DuPont Company's annual report contains information on the reduction in toxic air emissions and solid waste, as well as the amount of capital funds spent to improve environmental controls (DuPont Company, Wilmington, Delaware: Annual Report to Stockholders, May 2001).

The U.S. regulatory community, whether federal, state, or local, has also evolved significantly over the last 10 to 15 years. Whereas command and control approaches were the norm 10 years ago, and in many cases necessary to achieve the improvements in the environment that were needed, today there appears to be a greater shift towards a performance-based process where the regulated community is afforded the opportunity to meet environmental goals using their own approaches. Partnerships are being established among the regulated and regulators such that today's approaches to environmental improvements are becoming more collaborative in nature.

Nevertheless, what is now needed for the regulated and regulatory communities alike is a sound, science-based framework within which informed, reasoned, and transparent ecological risk management decisions can be made. With the publication of this book on risk management by SETAC, we hope that the framework and concepts exposed herein become widely used by policy makers in both the public and private sectors who are faced with making difficult environmental management decisions in the year 2001 and beyond. None of us (editors) intend to imply that we have developed the ultimate, everlasting paradigm for managing ecological risks in the U.S. or elsewhere. Neither do we advocate that the framework and concepts be viewed as a panacea for the environmental problems we face in the U.S.; there are more problems to address and this publication will not alleviate that need. The book is, moreover, an advancement of the dialog on how to manage ecological risks, a good road map for the discussions, a potential template for regulatory guidance, but certainly will require additional work and evolution as our experience base in managing ecological risks, and documenting them, improves. We also need to acknowledge that considerable information has existed on managing natural resources, similar to the information and approaches we present herein. We have attempted to capture some of that relevancy in our discussion by adding specific examples for managing migratory bird harvest and important habitats, yet did not make these the central focus of our work.

–Ralph G. Stahl, Jr.

CHAPTER 1

Introduction and Background to the Development of a Framework for Ecological Risk Management

Ralph G. Stahl Jr., Charles A. Pittinger, Anne L. Barton, James R. Clark, Peter L. deFur, Stephen J. Ells, Michael W. Slimak, Randall S. Wentsel, Richard B. Belzer, Robert L. Bachman

Purpose

This book is the product of a Society of Environmental Toxicology and Chemistry sponsored workshop on Framework for Ecological Risk Management held 23–25 June 1997 in Williamsburg, Virginia, as well as a compilation of discussions and debates held among risk assessors from the public and private sectors over nearly a 2-year period. It illustrates the complex issues associated with linking the science of ecological risk assessment (ERA) with the needs of the risk manager and decision-maker. It also defines and details a framework for ecological risk management (ERM). This framework will assist managers of ecological risks and natural resources in addressing ERA and ERM issues; it is not intended to prescribe standards for protection of ecological resources that can be highly contextual or mandated by particular laws.

We began the associated discussions and this book with three basic tenets:
- risk-based decision-making can be an important tool for managing ecological risks;
- there is a basic framework through which ERM decisions can be made in an efficient, timely, and scientifically supportable manner; and
- science can provide some, but not all, of the answers required for managing ecological risks.

Although there may be numerous applications of ERA and ERM, in this book, we have focused on waste sites, new chemicals and products, and natural resources. Obviously, there is some degree of overlap among these three applications and a much larger experience base in some applications than in others.

Even before the Workshop, our discussions led us to conclude that ecological risk managers must rely not only on the ERA, but also on input from social, economic, cultural, and scientific areas in order to successfully implement ERM actions. Lack

Risk Management: Ecological Risk-Based Decision-Making. Ralph G. Stahl, Jr. et al., editors.
©2001 Society of Environmental Toxicology and Chemistry (SETAC). ISBN 1-880611-26-0

of attention to any of these areas may result in an inappropriate decision. The book will not cover all of these areas in equal detail; however, as noted above, collectively the chapters will define and detail a process (framework) whereby these, and other, areas of consideration can be incorporated systematically into the decision-making process.

Many readers are likely to appreciate that the degree of risk to ecological resources deemed acceptable or unacceptable reflects a diversity of social and cultural values. Thus, it is unlikely that any one single approach to ERM will suffice in all possible situations. All of the chapters focus on issues commonly encountered in the U.S., since this allowed us a narrower focus through which we might reach consensus more easily. We hope that this book will stimulate similar debates and efforts outside the U.S., and we encourage our colleagues to engage us when appropriate.

From the frequent discussions among scientists and decision-makers, we identified three overarching ERM issues that ultimately drove plans for the Workshop and subsequent production of this book:

- What process is used (or should be used) to determine how we decide what ecological resources to protect?
- How do we decide to what degree we may (or should) protect those ecological resources?
- What tools are available for helping to make ERM decisions?

While there is nothing fundamentally different about making risk management decisions for humans as compared to ecological resources, both ERA and ERM are more complex than their human-oriented counterparts. First, ecological risk managers face unique challenges, including the diversity of species, scales of biological organization, and number of endpoints and criteria that might be considered. Human health-risk assessment (HHRA) focuses on a single species, humans, whereas ERA could potentially focus on important species numbering in the dozens or more. Second, ecological risk assessors and managers must contend with various levels of biological organization (individual, population, community, ecosystem, landscape) that are not major issues with HHRA. Third, throughout human history, society has developed and implemented processes and practices for measuring and managing risks to humans. Our historical knowledge of ecological resources is limited because we have not had an analogous effort for them during this same time period. Fourth, ecosystems do not "vote," and the difficulties in gaining public understanding of the issues make the challenge of risk management even greater. Finally, ecological management efforts often must balance purposeful exploitation or collateral impacts as ecological resources are utilized for the benefit of society against efforts to ensure sustainable ecosystem functions provided by those resources, which also benefit society directly or indirectly.

Although we have chosen not to review them here in detail, we recognize that there are a substantive number of research and management or decisional frameworks that attempt to distill some of the concepts offered herein for risk-based decision-making. We do briefly discuss those drafted by the American Society for Testing and Materials (ASTM) (ASTM 1997) and the Presidential/Congressional Commission on Risk Management (Presidential/Congressional Commission 1997) because there are some elements common to the ERM framework proposed herein and those proposed by these groups.

We also recognize that a wealth of information exists under the broad category of "environmental management" and that the proposed ERM framework is likely to contain some elements discussed by others under this heading. Our intent is not to duplicate these previous efforts, nor to dilute them, but to illustrate the uniqueness of the proposed ERM framework detailed in this book. We believe one of its unique features is that the proposed ERM is directly linked with the relatively well-established ERA framework developed by the U.S. Environmental Protection Agency (USEPA) (USEPA 1992). More importantly, our proposed framework is unique in that it begins with the central tenet that the ERA is a key element of the final risk management decision (USEPA 1995) and is no more or less important than the other considerations shown in Figure 1-1. Finally, we endeavored to use terminology in the ERM context that is familiar to practitioners of ERA, more so perhaps than to individuals who practice "environmental assessment and management" and who may not utilize or follow the ERA process or be familiar with its terminology.

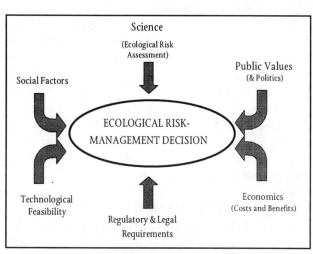

Figure 1-1 Inputs to the ecological risk management decision

Selected Terms and Definitions

Many terms are used throughout the book that may be unfamiliar to some readers. We discuss some key ones below.

At the Workshop we proposed a comprehensive definition for ERM:

Ecological risk management is the process of identifying, evaluating, selecting, and implementing cost-effective, integrated actions that manage risks to environmental systems while emphasizing scientific, social, economic, cultural, technological feasibility, political, and legal consider- . ations.

By this definition, ERM is a decision-making process that not only recognizes ecological risks in a conventional sense, but also incorporates an array of socioeconomic and other factors relevant to the decision. The best possible solution to decisions regarding chemical uses, industrial operations (including those led by government agencies), waste sites and natural resource utilization would be one that achieves a balance among all dimensions. Distinct "decision goals" can be articulated to reflect each dimension or factor, in terms compatible with each. These include attributes such as social acceptability, cost effectiveness, technological and logistical feasibility, reasonable durability, and adequate reliability.

Ecological risk managers

Ecological risk managers are individuals, groups, or organizations that use the results of the ERA along with economic, social, political, technological, and engineering inputs to decide a course of action and thus resolve environmental issues of concern. Risk managers could be decision-makers in regulatory agencies and businesses, as well as individual persons who must weigh variable information before making a decision on a particular issue. Many individuals, groups, or regulators may not have thought of themselves as "risk managers" in this broad context, but their actions to reduce potential ecological risks while maximizing other economic aspects or social benefits places them in a decision-making or risk management role.

Ecological resource managers

Ecological, or natural, resource managers are individuals or groups whose main focus is to ensure a sustainable supply of ecological resources such as important commercial or recreational plants or animals. Resource managers often are found in state agencies in which they are charged with setting harvest limits for particular plants or animals and in federal agencies that have the responsibility to manage natural resources for the national good. Ecological risk assessment may not be viewed by these individuals or groups as an approach to addressing their needs, yet it is clear that the concepts and actions involved with an ERA and risk management process are present albeit without the usual ERA "jargon."

Ecological risk assessors

Ecological risk assessors are the individuals or groups who collect and analyze scientific information and then use that information to characterize the risk to plants, animals, or habitats. They may also analyze this information and characterize the sustainability of plant and animal populations to aid ecological resource managers in making decisions.

Ecological resources

Ecological, or natural, resources are the plants, animals, and habitats that we are concerned about and want to protect. In ERA terms they might be known as receptors or ecological entities. They are the basis for the assessment endpoints that define what the ERA and ERM actions are all about.

Framework

Framework is a process guide for how to do something. It can take a number of forms and is generally flexible so that it can be applied to a variety of problems. A number of frameworks associated with ERA have been developed, and some will be highlighted in subsequent chapters.

Stressors

Stressors are physical, chemical, or biological "agents" that can pose risk to ecological resources when they are exposed to them. They may be natural or manmade, local or global, individual or multiple. Much of our experience in application is related to chemical stressors, generally because many of the federal environmental statutes are focused on regulating these stressor types.

Differences Between Ecological Risk Assessment and Ecological Risk Management

Generally, one of the hardest lessons that ecological risk assessors and environmental scientists have to learn is that ERM, and the decision-making process in general, is multifaceted and not wholly driven by scientific concerns. Ecological risk management decisions can appear to be inconsistent and controversial, which reflects, in large part, the wide diversity of inputs to the decision. In making an ERM decision, a decision-maker considers multiple issues: economics, social or political policy, technology, the "public," and science (e.g., the results of the ecological risk assessment). Some of the inputs to the decision are shown in Figure 1-1.

The "risk relevant" goal of ERM is to manage one or both of the two components of ecological risk—exposure and stress (hazard)— so that risks to ecological resources are controlled at acceptable levels or avoided altogether commensurate with other social or economic benefits that are being maximized. In general, the early stages of risk management are aimed at setting goals and identifying options, both of which serve to guide the subsequent risk assessment. This does not mean to set restrictions on the ERA, but simply to take into account important points and considerations when the ERA begins.

The product of an ERA is used by the ecological risk manager to understand the severity of the ecological risk, the ecological significance, uniqueness and integrity of the natural resources at risk, the ability of the system to compensate or recover, the availability of mitigation measures, the technologies available to mitigate the risk, the practicality in terms of cost, and other resources to mitigate the public's perception of the risk and the value dedicated to the resource. Other important factors, as illustrated in Figure 1-1, are involved:

- economics (costs and benefits)—cost-effectiveness of various risk-reduction or risk-mitigation alternatives and the benefits derived from each;
- social factors—public perceptions, risk communication, stakeholder involvement, environmental justice, and competing economic concerns (e.g., jobs, crime, education);
- regulatory and legal requirements—environmental laws and regulations; federal, state, and municipal policies; and associated processes by which they are developed;
- technological feasibility—availability of engineering capability to support various risk management options;
- public values—quality of life, clean environment, good health, good job, public perceptions, etc; and
- science (in the form of ERA)—the state of knowledge pertaining to the risk and its attendant uncertainty and variability.

In the context of the above, economics is the discipline that forces recognition of the value gained from protecting ecological resources and their comparison against the value of goods and services foregone. Social factors reflect public perception, risk communication, stakeholder involvement, environmental justice, and competing noneconomic concerns (e.g., public participation for its own sake, equity, and the distribution of the benefits and costs of ERM). This category is large and is often a difficult one to address given the conflicting representations of the public interest in any particular situation. Regulatory and legal requirements refer to current environmental laws and regulations, agency policy, and the actions that governments might take on behalf of the "environment." Technological feasibility encompasses the many facets of risk mitigation, the application of treatment options, pollution-prevention processes, life-cycle assessments, and others. Public values and "the public" are particularly nebulous terms because seldom are they interpreted consistently. Sometimes the term "stakeholder" is used synonymously with "the public," but this is not an appropriate usage (see Chapter 3). Often the public may value a particular ecological resource but may not be able to articulate the value in terms that are commonly used or measurable by scientists and public policymakers (McDaniels et al. 1995). Alternatively, there may be diverging values among various sectors of "the public" and no specific consensus presented to the risk manager. This presents a challenge for the ecological risk managers because they need to understand these values, or range of values, and communicate them to the ecological risk assessor. The ecological risk assessor is then challenged to collect and evaluate information relevant to those values and provide results of the ERA that address those values. If this communication fails, the results of the ERA and the possible ERM actions may be totally inappropriate (see Chapter 5).

Sound risk management decisions should reflect in large part the wide diversity of inputs to the decision (Figure 1-1). The vital policy–science interface requires that the process integrate social, political, economic, and technical interests and concerns. At the Workshop, it was proposed that scientific advances can shape or even dictate the development of public values and vice versa. An example discussed at the Workshop was the accelerated public debate and interest in policy development surrounding human gene manipulation, which became a controversial public issue only after the recent development of animal cloning technology.

Some risk management decisions are at times not wholly consistent with purely technical considerations of ecological safety or natural resource protection. Such decisions are often made when social, economic, or policy considerations are judged to take precedence. A specific example offered by a natural resource manager at the Workshop was the public outrage that accompanied and ultimately defeated a proposal for a hunting restriction on a depressed wildlife population in the state. In this case, the hunting tradition was valued and protected as a public "right," despite technical data that predicted the eventual demise of the population. While such decisions may be particularly disconcerting or misunderstood by the scientists who developed the ERA, it was recognized that ERM decisions often hinge upon prevailing public values or concerns. Likewise, ERM decisions based solely on technical considerations may not be fully accepted by stakeholders who may have their own set of unique values or concerns.

Ecological risk assessment is important to ERM because it is an appropriate analytical tool to help identify problems, compare the efficiency and effectiveness of ERM options, communicate to the public, and identify research needs. Because the purpose of sound environmental management is to protect human health and/or the environment, ERA helps to estimate, quantitatively and/or qualitatively, protection levels to achieve a particular outcome for the ecological resources being considered. The role of ERM in this process is to help the ecological risk assessor understand the decision being faced so that the appropriate information is developed. The risk-assessment process is also useful in the generation of health or environmental criteria used in environmental decision-making, an important element of risk-based decision-making. On the other hand, risk assessment alone typically will not provide a hard and fast target that can be applied to all situations, regulatory or otherwise.

The Parallel of Ecological Risk Management with Human Health Risk Management

In developing a framework for ERM, workshop participants frequently compared and contrasted human-health and ERA and ERM goals. Ecological risk management decisions are rarely independent of considerations of human health and are not always parallel. In instances in which ecological consequences may be severe, such as remediation efforts involving extensive soil removal and decontamination, balanced consideration must be given to both human-health and ecological goals as well as to the potentially adverse impact of the management action. Recent guidance developed by USEPA (Luftig 1998) illustrates this latter point.

Historically, human-health concerns have driven major environmental policies with respect to chemical usages and industrial operations. There appears to be broader consensus on "benchmarks" for HHRA endpoints (e.g., 1 in one million increased lifetime risk of contracting cancer), which are at times even embedded in statutory requirements. In contrast, ecological risks require consideration of many more endpoints for which there are fewer established benchmarks; hence, more decisions are made on a case-by-case basis. In ERM there is often greater reliance placed upon site-specific considerations and the professional judgment of the assessors. Invariably, these factors often give rise to significant controversy in ERM decisions.

Issues Associated with and Reasons for Making an Ecological Risk Management Decision

There are numerous reasons for conducting ERA and managing ecological risks, whether in the public or private sectors. In the U.S. several environmental statutes mandate protection of human health and the environment and have spawned national regulatory programs (Table 1-1). Some states augment these with state-specific programs, sometimes in response to the uniqueness of the ecological resources or unique regional or local values associated with the rescues.

Table 1-1 U.S. environmental statutes where ecological risk assessment is important

Statute	Common Name	Focus Area
Comprehensive Environmental Response, Compensation and Liability Act	Superfund (CERCLA)	Clean up hazardous waste sites.
Endangered Species Act	ESA	Protect plants and animals that are likely to be lost from the nation otherwise.
Federal Water Pollution Control Act	Clean Water Act	Regulate discharge of pollutants to waters of the nation.
Federal Insecticide, Fungicide and Rodenticide Act	FIFRA	Regulate new crop protection chemicals and agents.
Resource Conservation and Recovery Act	RCRA	Clean up active waste generating and storage facilities.
Toxic Substances Control Act	TSCA	Regulate non-crop protection chemicals and agents.

In the private sector, compliance with legislative and regulatory programs underlies a large portion of the ERA undertaken. However, a growing number of private sector concerns now use ERA and ERM in the evaluation of new or proposed products, disposition of corporate land, selection of emission control and waste-treatment technology, the siting of new facilities, and the prioritization of the resources expended on environmental issues. For example, a fundamental principle of Responsible Care®, the chemical industry's program for understanding and managing risks associated with the manufacture, transportation, use and disposal of products, is to develop the appropriate basic health and environmental data to make informed risk management decisions. Corporate policies also motivate this activity because it makes good business, economic, and environmental sense and stimulates more and more of the ERA conducted by the private sector (CMA 1997; AIHC 1998).

Risk managers make and communicate the risk-based decisions regarding what are acceptable levels of stresses imposed on ecological resources. The decision-making process can be facilitated by developing a list of associated options and questions under consideration and then describing the risks and consequences associated with each option. Many decisions may be contingent upon commitments to follow-up monitoring or supplemental studies to confirm assumptions or address residual uncertainties. Delaying a decision before such final data are collected, although often implemented, can be problematic for the affected parties and the public. Yet, few risk management decisions are made with "complete" datasets and little or no uncertainty. Hence, significant decisions are seldom made without generating some degree of controversy or criticism by one group or another.

At this juncture, it is important that the ecological risk assessors understand their roles in this process. The ERA is not the final answer but is one of several very important inputs to the overall risk-based decision process.

While separate Workshop sessions were convened to address the questions of how to decide what to protect and how to decide the degree of protection, there was consensus that the drivers, criteria, and decision process for addressing both questions are often synonymous. Identification of what to protect typically is accomplished at an early stage, while the degree of protection is determined through collective consideration of the drivers noted above. It was agreed that the context of the ecological risk often determines, to a large part, what to protect and the degree of protection deemed acceptable. A distinction was made between "crisis risk management" (e.g., a derailment spill) from "systematic risk management" (e.g., a new chemical or pesticide evaluation, a natural resource management plan). Workshop participants viewed different processes and criteria to be operational in each case.

Equally important in deciding the degree of protection can be the affiliation of the risk manager, whether in a public (regulatory) or private (corporate) capacity. Both must satisfy multiple stakeholder groups, albeit differing in composition and often differing in priorities. As such, criteria and procedures for determining "acceptable risk" typically vary. The regulatory risk manager must serve the expectations and concerns of the public, the regulated community, and elected legislative bodies in a highly visible fashion often dictated by prescriptive statutes. Divergent concerns and sometimes inconsistent demands surrounding the regulatory manager favor a cautious approach, with few incentives and little reward for innovation and flexibility. Legislative edicts may require agencies to adopt a "tunnel vision approach" to risk management. For example, single medium statutes (e.g., the Clean Water Act) may at times restrict considerations of potential risks in other media.

Routine environmental risk management decisions by corporations (e.g., in the formulation of new products, the development of manufacturing processes and the management of industrial operations and remediation of contaminated sites) are usually less subject to exhaustive public scrutiny than are decisions made by public officials. Nevertheless, the private risk manager also answers to an array of stakeholders. Unique demands are placed upon private risk managers by internal management lines that ultimately reflect stockholders' expectations of financial success and sound business practices (see Chapter 10). National and international regulatory conventions and standards impose criteria for decision-making. Equally important is how the corporation's risk management decisions are viewed by the consuming public. Hence, private sector risk managers are challenged to satisfy a set of fundamental environmental management needs including human and environmental safety, regulatory compliance, efficient resource use and waste management, and satisfaction of public (societal) concerns. The first two elements are obligatory prerequisites for corporations to remain in operation, and ideally should not be compromised by commercial or other objectives. The latter two elements are less tangibly related to conventional risk assessment than to corporate profitability and the long-term success of the business but must nevertheless be weighed in management decisions. Innovative solutions that simultaneously accomplish each of these goals are encouraged, particularly in an increasingly global and visible business atmosphere.

Risk management approaches in the U.S. historically have focused on the protection of individual species, and even individual animals, threatened by isolated risks. For example, Cairns and Pratt (1993) illustrate the use of benthic macroinvertebrate biomonitoring as a tool for decision-making and in the process detail how the early focus of this work was to provide information at the individual level. Recently, more attention is being given to consideration and protection of higher levels of biological organization, including populations (Luftig 1998), communities, and ecosystems (e.g., the Florida Everglades). Current methods to assess exposure and response at these higher scales of complexity with acceptable levels of precision are generally lacking. Similarly, greater emphasis needs to be placed on assessment of multiple stressors acting simultaneously in time and space (e.g., the net effects of physical, chemical, and/or biological stressors) and on more holistic perspectives (e.g., on community-and ecosystem-level endpoints) in risk management decisions (Foran and Ferenc 1999).

Statutory criteria and discretionary regulatory policies of federal agencies in some cases narrowly define the natural resources to be protected (see Chapters 4, 7, and 8). The Endangered Species Act and Marine Mammal Protection Act are among the more prescriptive environmental statues that explicitly dictate protection measures for individual species. The Toxic Substances Control Act and the Comprehensive Environmental Response, Compensation, and Liabilty Act (CERCLA) tend to be less prescriptive in this regard, but can be equally contentious. Statutory and regulatory risk management policies or standards were viewed as either discretionary (e.g., the aesthetic or recreational value of a natural resource) or non-discretionary (e.g., under CERCLA, any applicable, relevant and appropriate requirements invoked in a particular decision). In either case, even statutory criteria were seen as the expression of a public value at a point in time (i.e., the passage of the enabling legislation) and are subject to change as a result of evolving political considerations and public sentiments.

In a majority of statutes, protection goals for key natural resources (e.g., the habitat of an endangered species versus the organism itself) are not explicitly stated (see Chapter 4), providing ample room for interpretation by industrial organizations and agencies with enforcement authority. This, too, has amplified the controversy of environmental management decisions in the U.S., often resulting in civil or criminal litigation. The degree of protection to be afforded commonly is defined vaguely in qualitative terms that leave considerable discretion to interpretation, although there may be instances in which precedents are followed or benchmarks have been established that serve as guides. Current land-use and reasonable future land-use scenarios were identified at the Workshop as important to decisions involving waste-site remediation and often are used as guides in final decisions on cleanups. In addition, natural resource agencies and conservation groups (see Chapters 8 and 13) have either explicit or implicit criteria and benchmarks that are used in decision-making, allowing for interpretation based on a particular situation.

Some at the Workshop voiced concern that natural resource attributes that traditionally are valued by the general public (e.g., abundant fishing and hunting, recreational opportunities, scenic views) do not reflect sufficiently the requirements of well-functioning ecosystems. At times the public may appreciate the value of the natural resource as a whole, yet may overlook the ecological significance of a given species or community. In other cases, there was concern that the public may genuinely value a particular ecological resource (e.g., wetlands), yet be unable to articulate its value in quantitative measures meaningful to scientists or actionable by policymakers (Dover and Golding 1995). It was noted that contingent valuation techniques (to estimate economic worth through comparisons with known commodities) and other controversial procedures are being increasingly developed and applied by natural resource economists to ascertain the public values placed upon natural resources or aesthetic and recreational qualities of the environment. The need for more research and validation in these areas was emphasized. An overview of economic considerations and their application in ERM is given in Chapter 6.

A challenge for the ecological risk manager is to understand public values and communicate them to the ecological risk assessor. The risk assessor is then challenged to select ecologically meaningful endpoints and to collect and evaluate information relevant to those values. Ecological risk management goals may be viewed as representing operationalized public values expressed in measurable and, hence, manageable terms. Communication with the lay public by risk assessors and risk managers is critical in gaining support for protecting little known species, ecological communities, or potentially undervalued habitats.

Settings and Contexts of Ecological Risk Management

Many risk management decisions are associated with a specific geographic location (e.g., a hazardous waste site, release of an effluent into a local stream, acreage of habitat threatened by a proposed development, etc.), commonly termed place- or community-based assessments. In this case, ecological risk managers are those who are responsible for the property or activity posing the ecological risks, those who are charged with protecting the ecological health of adjacent properties or receiving systems, or those who issue permits for an activity. Other ERM decisions address very broad perspectives and deal with issues common to a generalized or hypothetical ecological community that is representative of the variety that might be encountered across broad geographic scales (e.g., forests, wetlands, lakes, or rivers) and are focused on common environmental media or representative biological communities. Here the decision process usually concentrates on prospective assessments of potential impacts to ecological resources in soil, water, or sediments and seeks actions that will be broadly protective of resources in most locales. For these decisions, ecological risk managers are those who propose to undertake the activity (e.g., new construction processes for facilities, power lines, or roads; proposal of the use of a new chemical; introduction of a new plant species to control soil erosion on steep hillsides) or those who are charged with protecting environmental health with respect to such activities. Some ERM decisions may focus on large-scale, but well-defined, regional resources within a specific watershed (Delaware River Watershed) or jurisdictional boundary (U.S. portions of the Great Lakes).

For the purposes of the Workshop discussions, ecological risk managers were broadly defined as individuals, groups, or organizations that apply the results of an ERA or other resource assessment tool, along with economic, social, political, technological, and engineering inputs, to determine a course of action to resolve an environmental issue. Ecological risk managers encompass both public (e.g., federal, state, or municipal regulatory officials or legislators) and private (e.g., corporate landholders or conservation groups) affiliations. Beyond the management of anthropogenic risks, there was consensus that the ERM concept equally applies to natural resource managers with responsibility for conserving, protecting, or nurturing important commercial or recreational plants, animals, and ecosystems. While natural resource managers (e.g., federal and state wildlife conservation officials) traditionally may not identify themselves and their work with "ecological risk assessment" per se, it was recognized that resource management goals and criteria are often identical to those of conventional "risk scientists." The resource managers at the Workshop were thus able to contribute a valuable and practical perspective to discussions on decision-making criteria.

The National Research Council's (NRC) report "Understanding Risk: Informing Decisions in a Democratic Society" (NRC 1996) describes the importance of interested and affected parties in the risk-assessment process. The report stressed that risk assessment should be a "decision-driven activity" directed toward informing choices and solving problems. It is the outcome of an "analytic-deliberative process," in which "analysis" uses methods to arrive at answers to factual questions and "deliberation" is the process of communication and collective consideration of values and issues. To begin to answer the "what to protect?" question that has challenged ERA practitioners, analytic-deliberative methods must be developed and used at all stages of an ERA, not only by risk managers but also by stakeholders. Additional information on risk communication is provided in Chapter 5.

Selection of appropriate stakeholders can be problematic: Who should have a say? Are important perspectives not represented at the table? Stakeholder input is important because value judgments must be taken into account by risk managers (Chapter 6). The socioeconomic relationship of stakeholder groups to the risk assessment issues and to each other is critical to the outcome. Deliberate consideration of stakeholder involvement will lead to better risk management decisions. The appropriate blend of stakeholders and the extent and type of their involvement will vary from step to step, as well as from setting to setting. In addition, ecological risk decisions frequently deal with significant uncertainties that should be conveyed in meaningful terms. Workshop participants emphasized the need for better risk communication to the lay public, recognizing that the more localized the issue is, the greater the need for transparency is.

Frameworks

Defined earlier, frameworks are process guides for how to do something, and we have chosen not to review all of them in great detail. In the context of ERA however, several such frameworks have been developed, beginning with the USEPA Framework for ERA (USEPA 1992). Subsequent efforts have included the Aquatic Dialogue Group (Baker et al. 1994), Environment Canada (Gaudet 1994; CCME 1996), European Commission (EC 1994), Nordic Council of Ministers (Pedersen et al.

1994), U.S. Tri-Services (Wentsel et al. 1996), United Kingdom (UK Department of the Environment 1993), Water Pollution Control Federation (Parkhurst et al. 1990), and perhaps many others.

Currently in the U.S. there is no nationally recognized framework for ERM. In 1983, the NRC's "Red Book" described the distinction between risk assessment and risk management. It did not distinguish human-health applications from ecological applications. Other groups, such as ASTM (1997), recently drafted an ERM framework that did not find a high level of interest from either the public or private sector. In early 1997, the Presidential/Congressional Commission on Risk Assessment and Risk Management released a framework for risk management that is directed primarily at managing risk to human health (Presidential/Congressional Commission 1997). This latter document provided only a cursory discussion of ERM.

The ecological risk assessment framework of the U.S. Environmental Protection Agency

In the U.S. the basic framework for ERA (framework) was developed in the late 1980s and was released formally as USEPA guidance in 1992 (USEPA 1992). The Framework by design provides a broad outline of the major steps involved in ERA and provides general guidance on what should occur in each of these steps. Other groups, including the USEPA Program Offices and several states (Massachusetts, Pennsylvania, Texas, and Washington), are developing more detailed guidance on ERA to supplement the general guidance provided by the Framework. It generally is accepted that the Framework has gained wide utility both in the public and private sectors, whether addressing waste sites or new or existing substances.

The USEPA's current ERA process was developed with very broad stakeholder input and review. It is depicted in Figure 1-2 taken from the recent guidelines for implementing the framework (USEPA 1998).

The USEPA guidelines effectively divide ecological risk assessor activities into four steps: 1) planning, 2) problem formulation, 3) analysis, and 4) risk characterization. These four steps in the ERA process subsequently provide input to a risk management decision. Risk management is separated from the ERA because the process is not driven by science alone. Yet, input into risk management could be considered as the fifth and final step in the overall process of risk assessment and risk management. To maximize the efficiency and integrity of the decision-making process, it is important that the separate and individual roles and responsibilities of risk assessors and risk managers be understood and agreed to before the process of making a risk-based decision is begun. While the USEPA guideline document highlights the role of the risk assessor in each step of the risk assessment process, the role of the risk manager has been defined less specifically. In Table 1-2, suggested roles are that the ecological risk manager may play in each of these four steps.

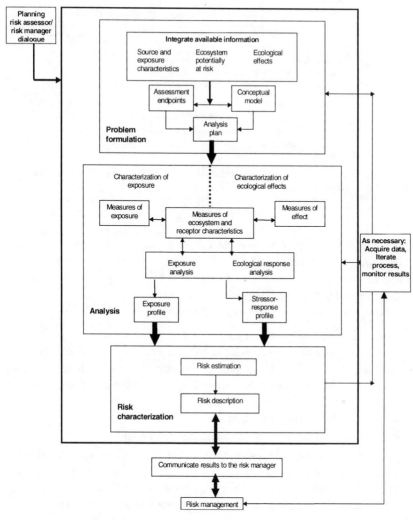

Figure 1-2 The USEPA framework for ecological risk assessment (1998) (redrawn).

Planning the ecological risk assessment

To ensure that ecological risk managers interested in this topic are familiar with the ERA process, the following discussions are presented to lead them through the various steps in an ERA. The first step, and one in which the ecological risk manager plays a major role, is planning the assessment.

The potential complexities of an ERA demand careful planning for its design. Many of the complexities differ from those of HHRA and include deciding what species, populations, ecosystems, or functions are most relevant. Ecological risk assessment also should address species interactions and indirect effects, as well as the signifi-

Table 1-2 Possible roles of the ecological risk manager in the ecological risk assessment process

Step in the ecological risk assessment	Role of the ecological risk manager	Level of ecological risk manager involvement
1. Planning	Help develop scope, timing, description of regulatory and resource constraints, and overview of ecological risk management issues. Identify broad policy goals and objectives.	High
2. Problem formulation	Help establish assessment and measurement endpoints, provide additional background information.	Moderate / High
3. Analysis	Communicate with ecological risk assessor on data collection issues, changes in risk management goals, help with resource problems.	Low
4. Risk characterization	Understand the type and magnitude of ecological risk posed by stressors and uncertainties in the assessment.	Moderate / High

cance of the three stressor types: biological, chemical, and physical. Environmental assessments must be conducted within the bounds of the financial, logistical, and temporal limitations surrounding an issue, which will greatly affect the magnitude of effort contributed to the ERA. Depending on the scope and detail of the effort, some or all three stressor types may be evaluated simultaneously.

The most important part of the risk manager's job in planning is to set specific objectives and articulate the questions for risk management. As a part of this, the risk manager needs to involve affected parties, including the public; however, public involvement may or may not be needed depending on the magnitude and scope of the potential problem being addressed. Some specific risk management tasks that should be accomplished during planning are addressed below.

1) Identify the problem (why we are doing this assessment).

2) Determine the ecological resources relevant to the problem and the valued ecological resources that need to be managed (for a value-driven assessment). Determine what is in the area (for a place-based assessment) or what is vulnerable to the chemical, biological, or physical stressor (for a stressor-based assessment).

3) Bring in the affected parties, the public (where needed), and the ecological risk assessors. The public's major role is to help determine the societal value of the resources potentially at risk, especially when arrayed against alternative ecological resources that could be protected instead and against goods and services that society must forego in order to protect the ecological resources in question. The questions listed here can serve as a guide to help the risk manager frame the problem. (It is also useful to include the assessors or other scientists at this stage so that the connection between societal value and ecological relevance can be discussed. This is the point at which it is important to discuss the time and resources that can be allocated to the effort, as these factors often dictate the reasonable scope of the objectives.)

4) Set objectives. The objectives will be based upon the information gained in Step 3. Objectives provide good guidance for the rest of the process if they specifically include the resource to be protected, the aspect of that resource to be protected, and the desired state. An example would be to reestablish a self-sustaining scallop population that can support a viable fishery.

In addition to the questions, the risk manager should determine how the public and affected parties will be represented, how their input will be obtained, and how that input will be utilized in the planning process or at other points in the assessment.

A final step in the planning phase is for the ecological risk manager to recheck the level of effort required in the assessment, the timing involved, and other logistical issues associated with its conduct to ensure that allocated resources are commensurate with the study objectives and expectations of the stakeholders. As illustrated previously above, thoughtful questions and early planning are necessary to ensure that ERAs are conducted efficiently and thoroughly, providing an outcome relevant to the decision at hand.

Problem formulation

Problem formulation is the step at which ecological risk assessors develop a conceptual model of how the selected ecological resources are exposed to and affected by the stressors under consideration. Throughout the assessment the role of the ecological risk manager is to ensure that the efforts stay focused on getting the information needed to make a risk-based decision and that the assessment endpoints reflect the management objectives for the ecological resources anticipated to be at risk.

In the problem-formulation step, risk assessors characterize the issues and resources under consideration by using existing and available data to organize the ERA efforts. The problem-formulation step provides a clearer understanding of the direction and effort needed to complete the ERA and often leads to more discussions with the ecological risk manager regarding complexity of the issues, resources to be committed, costs, and timing. Major assumptions and uncertainties that may or may not be addressed, specifically during planning or at other points in the ERA, are identified and discussed during this step. Other activities in problem formulation include the development of an analysis plan to guide the collection and interpretation of data relevant to the ERA.

Analysis

Environmental risk assessment focuses on the technical aspects of exposure and effects for the ecological resources under consideration and the economic valuation of such resources. If needed, development of quantitative estimates of exposure via various pathways for ecological resources exposed to a stressor may involve a variety of measurement and analytical tools. Empirical data from laboratory studies or field-monitoring programs may be used along with computer simulations of chemical transportation and fate in the environment. Effects analyses focus on characterization of exposure–response relationships and evaluations of the spatial and temporal scale of these relationships. In risk analysis, the ecological risk assessor quantifies the probability of observing small to severe effects, given the various exposure patterns identified. Economic analysis measures society's willingness to pay to protect or enhance the quality of an ecological resource. Quantification of uncertainty associated with these evaluations is a key attribute for these analyses, as the final decision must incorporate this factor.

In practice, all these aspects of the analysis phase may be completed in a tiered fashion. As the ERA in general, and the analysis phase in particular, moves from lower tiers to higher tiers, these efforts move from qualitative to quantitative analyses. The goal is to achieve the level of certainty in the analysis that is commensurate with the significance of the risk issue at hand and the availability of data and resources to advance the analysis.

Ecological risk managers typically have a very limited role in the analysis step, because of concerns over the potential to influence the outcome. Management involvement in this step could be interpreted as an attempt to bias the analyses or assumptions in order to direct the outcome of the ERA towards a self-serving decision. Risk-assessment practices have been guided by institutions such as the NRC (NRC 1996, 1983) and the Presidential/Congressional Risk Commission (Presidential /Congressional Risk Commission 1997). In general, management involvement is low during this phase, although risk managers may have a role when risk assessors develop difficulties in executing the analysis because of limitations in resources or guidance Also, it is important that risk managers understand the assumptions and uncertainties inherent in the analyses to ensure that they will comprehend the value of the outcome.

Risk characterization

Risk characterization is the integration of exposure and ecological effect information and the development of explicit expressions regarding the risk to the selected ecological resources. During risk characterization, ecological risk managers ensure that the information is expressed in a manner relevant to the questions at hand.

Because the risk-based decision may have considered, but not necessarily agreed with, the competing interests of all stakeholders, the ecological risk manager should have a firm understanding of the uncertainties associated with the expression of risk, the assumptions that went into the analyses, and the limitations of the technical information. Also, risk characterization should address spatial and temporal scales of the ecological risks under consideration. Ecological risks should be characterized in terms of probability of occurrence and magnitude of effect. They should address the likelihood and rate of ecological recovery if impacts do occur and place the risks into a context relative to other similar or known ecological risks for comparison. The risk manager should understand these issues, and their uncertainties and limitations, in order to factor them into a risk management decision.

Structure of the Book

The book is structured into two main sections and 15 chapters. Section 1 (Chapters 1 through 6) contains those chapters specific to the elements of the ecological risk management framework, as well as supportive information. Section 2 (Chapters 7 through 15) contains chapters that illustrate the application of ecological risk management in various contexts, as well as an invited chapter on the perspective of what might be needed in ecological risk management. Also in Section 2 is a concluding chapter that details the recommendations and general conclusions from the Workshop.

References

[AIHC] American Industrial Health Council. 1998. Ecological risk assessment. Sound science makes good business sense. Washington DC: American Industrial Health Council. 13 p.

[ASTM] American Society for Testing and Materials. 1997. Proposed guideline for ecological risk management. Conshohocken PA: ASTM.

Baker J, Barefoot A, Beasley L, Burns L, Caulkins P, Clark J, Feulner R, Giesy J, Graney R, Griggs R. 1994. Aquatic dialogue group: Pesticide risk assessment and mitigation. Pensacola FL: SETAC. 220 p.

Cairns J and Pratt JR. 1993. A history of biological monitoring using benthic macroinvertebrates. In: Rosenberg DM, Resh VH, editors. Freshwater biomonitoring and benthic macroinvertebrates. New York NY: Chapman and Hall. p 10–27.

[CCME] Canadian Council of Ministers of the Environment. 1996. A framework for ecological risk assessment: General guidance. Winnipeg Manitoba, Canada: CCME. 32 p.

[CMA] Chemical Manufacturers Association. 1997. Ecological risk assessment: A tool for decision making. Arlington VA: CMA. 20 p.

Dover MJ, Golding D. 1995. Communicating with the public on ecological issues: workshop report. Worcester MA: Clark University, Center for Technology, Environment and Development. 47 p.

[EC] European Commission. 1994. Risk assessment of existing substances. Technical guidance document. Brussels, Belgium: European Commission, Directorate-General, Environment, Nuclear Safety and Civil Protection. XI/919/94-EN.

Foran JA, Ferenc SA, editors. 1999. Multiple stressors in ecological risk and impact assessment. Pensacola FL: SETAC. 100 p.

Gaudet C. 1994. A framework for ecological risk assessment at contaminated sites in Canada: Review and recommendations. Ottawa Ontario Canada: Environment Canada, Ecosystem Conservation Directorate. Scientific series Nr 199. 108 p.

Luftig S. 1998. Draft ecological risk management principles for Superfund sites. Washington DC: OSWER Directive 9285.7. 28 p.

McDaniels T, Axelrod LJ, Slovic P. 1995. Characterizing perception of ecological risk. *Risk Anal* 15(3):575–588.

[NRC] National Research Council. 1983. Risk assessment in the federal government: Managing the process. Washington DC: National Academy Press.

[NRC] National Research Council. 1996. Understanding risk: Informing decisions in a democratic society. Washington DC: National Academy Press.

Parkhurst BR, Bergman HL, Marcus MD, Creager CS, Warren-Hicks W, Olem H, Boelter AM, Baker JP. 1990. Evaluation of protocols for aquatic ecological risk assessment and risk management. Alexandria VA: Water Pollution Control Federation Research Foundation.

Pedersen F, Leinonen R, Pettersson I, Norrthon S, Kalqvist T, Hansen H, Gjos N, Haegh GS, Lander L, Petersen G, Dahlbo K, Kemilainen B, Wenell T. 1994. Environmental hazard classification—classification of selected substances as dangerous for the environment. Copenhagen, Denmark: Nordic Council of Ministers, TemaNord 1994. 101 p.

Presidential/Congressional Commission on Risk Assessment and Risk Management. 1997. Final Report. Volume 1. Washington DC. 64 p.

UK Department of the Environment. 1993. Risk assessment of existing substances. London, England: The UK Government/Industry Working Group. 69 p.

[USEPA] U.S. Environmental Protection Agency. 1992. Framework for ecological risk assessment. Washington DC: USEPA. EPA-630-R-92-001.

[USEPA] U.S. Environmental Protection Agency. 1995. Ecological risk: A primer for risk managers. Washington DC: USEPA. EPA-734-R-95-001.

[USEPA] U.S. Environmental Protection Agency. 1998. Guidelines for ecological risk assessment. Washington DC: USEPA. EPA-630-R-95-001F.

Wentsel RS, LaPoint TW, Simini M, Ludwig D, Brewer L. 1996. Tri-service procedural guidelines for ecological risk assessments. Volume I. Fort Belvoir VA: Defense Technical Information Center. Document ADA314323. 108 p.

CHAPTER 2

A Multi-Stakeholder Ecological Risk Management Framework

Charles A. Pittinger, Jessica Glicken Turnley, Paul M. Mehrle

This chapter provides an overview of the ecological risk management (ERM) framework developed at the Society of Environmental Toxicology and Chemistry (SETAC) on Framework for Ecological Risk Management Workshop 23-25 June 1997 in Willamsburg, VA described in Chapter 1 framework and its complement. The framework consists of seven iterative elements (Figure 2-1) as described in the Workshop Summary (Pittinger et al. 1998) and is complementary to the U.S. Environmental Protection Agency's (USEPA) Framework for Ecological Risk Assessment (ERA) (USEPA 1992). Subsequent chapters of this book elaborate on the various elements of the framework and illustrate how it is applied practically in public and private decision-making processes.

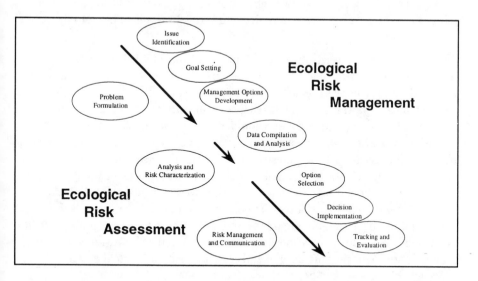

Figure 2-1 Relationship of the ecological risk management framework (Pittinger et al. 1998) to the ecological risk assessment framework (USEPA 1998)

Risk Management: Ecological Risk-Based Decision-Making. Ralph G. Stahl, Jr. et al., editors.
©2001 Society of Environmental Toxicology and Chemistry (SETAC). ISBN 1-880611-26-0

Linking Ecological Risk Assessment and Risk Management Frameworks

In the USEPA's ERA Framework (Chapter 1), the association with risk management is acknowledged in a planning discussion between the risk assessor and risk manager before the assessment is carried out and in another discussion following the assessment to link the results to the decision at hand. It was not the intent of those in the Agency who described the ERA Framework or subsequently developed the Agency's Guidelines for ERA (USEPA 1998) to elucidate the nature of these discussions or the management process they concern. Many believe that additional guidance for ERM would be a useful complement to the current assessment guidelines.

The risk management framework developed at the SETAC workshop is a conceptual umbrella linking ERA and ERM functions. It is consistent with the USEPA ERA Framework because it recognizes management needs (i.e., discussions between the assessor and manager) that ideally occur before, during, and after the technical risk-assessment process. These are linked through a feedback loop similar to that existing in the risk-assessment model, a mechanism presented in the assessment model as "data acquisition, verification, and monitoring." In the management of ecological risks, however, the "data" and "monitoring" may include diverse disciplines such as regulatory law, economics, and social norms in addition to the scientific context.

The quantitative and descriptive science used to conduct ERA does not answer, in a direct way, the question of what should be done to manage the risk. Science determines adversity, whereas the public determines acceptability. The key difference between the risk assessment and its reliance on quantitative empiricism, and the normative science used in social decision-making, is that a risk assessment will not convey the value of natural resources to stakeholders and decision-makers. To be objective, the risk assessor should strive to minimize the introduction of social and economic values into the assessment. Certain elements of the assessment process such as the selection of assessment endpoints, however, are invariably linked to the manner in which natural resources are perceived and valued. For example, soil concentrations of certain pollutants that might be subject to dermal uptake would be of high concern to Native American communities in the Southwest where pot-making is still an important cultural and economic activity, while they would be of much less interest to neighboring Anglo communities. In these circumstances, the risk assessor should clearly identify ("make transparent") the value-laden questions and policy decisions made. The use of normative analysis can never be wholly scientific (although it should be highly rigorous) because it involves values that are rooted in human experience. The environmental risk manager thus must be prepared to acquire and evaluate information from legal, political, cultural, social, economic, and technological studies. The conduct of these studies should be as professionally managed as is the ERA. As a result, resources for these types of studies need to be included in an environmental risk management budget.

The first stage of the management process, consisting of problem identification, goal-setting, and management options development, can be identified as the process occurring prior to and during the "risk assessor/risk manager dialog box" of the USEPA's ERA process (Chapter 1). The data compilation element of the management framework corresponds to the conventional ERA process (e.g., problem formulation, effects assessment, exposure assessment, and risk characterization). The last three elements of the ERM framework—option selection, decision implementation, and tracking/evaluation-represent—the "risk management" and "risk communication" elements of the ERA process.

This framework builds upon other risk management frameworks that have been proposed (see Power and McCarty 1998) because many of the concepts and processes of these frameworks, primarily derived from consideration of human-health risk, are fundamentally relevant and applicable to ecological risks as well. In 1983, the National Research Council's (NRC's) "Red Book" (NRC 1983) described the distinction between risk assessment and risk management; it did not distinguish human-health and ecological risk applications. In 1997, the Presidential/Congressional Commission on Risk Assessment and Risk Management (Presidential/Congressional Commission 1997) published a risk management framework based upon extensive dialogues with various stakeholder groups across the U.S. The American Society for Testing and Materials (ASTM 1997) also has initiated the development of an environmental risk management framework. Each has emphasized the role of stakeholder involvement as a means to satisfy the public values that are paramount in a management decision.

As discussed throughout this book, and particularly in Chapters 3 and 4, broad, multi-stakeholder communication is central to the framework as envisioned by participants at the Williamsburg workshop. In fact, the communication among stakeholders and between risk assessors and risk managers was viewed as a pervasive part of the risk management process, starting from the first step of identifying a problem through the final step of implementing a decision. The idea to explicitly represent "communication" as a discrete step was discussed, but was rejected on the grounds that it would seem to limit its importance or restrict it to a single stage. Rather, effective communication must occur throughout the decision-making process.

Framework Elements

The ERM framework from the SETAC workshop is multidisciplinary because it recognizes the importance of balancing multiple considerations in a management decision, e.g., scientific risk assessment, analyses of regulatory and legal requirements, political issues, social factors, environmental economics, and cost-benefit issues. The framework is suitable for numerous environmental risk management applications and is therefore adaptable to issues related to contaminated sites, product development, natural resource management, waste management, and other ecological concerns. For all applications, stakeholder involvement and the development and use of science-based information are encouraged. The following discussion of each stage underscores the involvement of multiple disciplines and stakeholders in managing ecological risks.

Issue identification

The three elements issue identification, goal-setting, and management options development typically occur in unison in an ecological risk management process. They are complementary to the "problem formulation" element of the USEPA's framework for ERA (Chapter 1). Issues are generated through a number of mechanisms, including commercial or government activities, legislative requirements, public concerns, advocacy by interest groups, the dissemination of new information, or the availability of new technology. The way that an issue may arise frequently determines its course of management and the degree of urgency given to its resolution. This will be discussed further in Chapter 4.

Goal-setting

During goal-setting, the environmental issue is translated into a series of goals and objectives conducive to systematic analysis and planning. Environmental protection goals in a democratic society are established in a variety of ways and may involve a large number of stakeholders, e.g., legislators and the constituencies that elect them; natural resource managers and regulators at the federal, state, and local levels; public planners, economists, local business, and national trade associations; and non-governmental organizations, academia, and the media. Likewise, they are expressed through a variety of means including formal executive orders, regulatory rules, and legislation. These goals also arise from case law and other judicial processes. Finally, beyond the formal mechanisms through which government business institutions operate, environmental goals emerge as a ground swell of popular public opinion in reaction to issues (e.g., competing needs for funding for education or crime) and events (e.g., catastrophic accidents such as spills or forest fires). Given this diversity, it is not a small task for the risk managers to recognize and articulate the environmental goals and to apply them to the issue or decision at hand.

Goal-setting must recognize not only ecological goals but also economic, human-health, political, legal, and social objectives. It challenges risk managers to frame the issue in meaningful, measurable, and achievable terms in order to define subsequent data-compilation needs. Values of diverse interests are often in competition. For example, commercial waterfront property development may compete with the need to maintain an undisturbed wetland habitat that supports aquatic communities. Competing and complementary interests should be identified and stakeholder representatives should be brought to the table early in the process in order to ensure maximum acceptability of the analysts' results and an ultimate management decision.

How is the goal articulated? Ideally, it is captured explicitly in a concise statement of the issue that addresses all stakeholder concerns and is accompanied by a statement of goals. The goals define what to protect by specifying the valued natural resources (e.g., a stretch of river), key attributes of the resources (e.g., the fish community, waterfowl habitat), and the desired state or outcome of the management actions (e.g., safe consumption of fish, breeding and rearing success of waterfowl). Each goal can be implemented by specific objectives guiding subsequent actions. Such an approach enables risk assessors to tailor their selection of assessment endpoints in a way that maximizes their relevance and utility for decision-making. The increasing interest and importance, accorded to active stakeholder participation in the definition of goals, is treated in Chapter 4 and is an integral part of this risk management framework.

Management options development

This element of the ERM process recommends explicit identification of apparent management options based on readily available, possibly qualitative, knowledge. It further recommends that the risk managers establish an initial set of decision criteria by which the tradeoffs among various options will be made. Both the management options and the decision criteria identified prior to data compilation and analysis should be viewed as flexible, accommodating revision or expansion as new options or concerns are identified.

The key to successful risk management is the identification of options that will lead to decisions supporting "acceptable" risk (which may at times represent no risk at all). Rarely is a risk assessment conducted without some prior consideration being given to the range of potential management options available to the risk managers and stakeholders. Understanding the extent to which management options are known is valuable in helping to tailor empirical studies that support the ultimate decision to be made.

"Acceptable risk" is a highly subjective and relative term. It is time- and space-specific and depends upon definitions of quality of life and robustness of the environment. Clearly, this is an exercise in tradeoffs. Stakeholder input is critical to identifying and defining the criteria by which these tradeoffs will be made (the "metrics" and their values for defining environmental robustness) and the relative weights that need to be given to the criteria. This input should be considered objectively by the risk manager to inform the decision as to which option, or combination of options, offers the best single solution to all collective criteria.

As recognized by the NRC (NRC 1983, 1996), the nature of the interaction between risk assessor and risk manager must be balanced to ensure objectivity in the assessment. Ecological risk assessments should not be "force-fitted" to justify preordained management decisions. If the scope of the assessment is constrained by the prerequisites of given management options, the process may be perceived as biased. At the same time, an uninformed risk assessor with little knowledge of the management options or purpose of the assessment may fail to produce relevant and decisive information. The risk manager should not unduly influence the risk assessor with regard to a particular management option but should inform the assessor of the technical feasibility of identified options and of important stakeholder issues. Only in this way can meaningful assessment endpoints be chosen.

The assessment process enables the risk manager to select the option best suited to the technical, regulatory, economic, and political constraints of the decision. Options may range across a broad spectrum extending from "no action," to "monitor the situation," to "limited management," to "virtual risk avoidance." Initially the list of options may be extensive, and may include those that may not at first glance be deemed feasible or practical. Options may subsequently be sorted into categories based on like properties or outcomes.

It is important that the management options identified result in a positive or no-net-loss environmental action so that those actions do not cause more harm to the environment than does the risk that is the focus of the management effort, unless social, legal, economic, or other factors are deemed to override environmental considerations. It is conceivable, for example, that input from a community regarding the economic consequences of a factory shutdown may cause a risk manager to seek other options (assuming that the factory meets emissions standards and other regulatory requirements), although environmental damage from continued operation may be significant.

Parallel to the development of a list of practicable management options, and equally dependent upon broad stakeholder input, is the establishment of clear decision criteria. The criteria should reflect the key concerns of stakeholders and the main drivers of the decision. A variety of decision-analysis methods are available for identifying and weighting criteria in risk- management decisions (see Engi and Glicken 1995; Saaty 1990). At times, the subsequent risk- assessment process identifies unique considerations of the risk or the ecological entity to be protected. These considerations may alter the initial concerns voiced by stakeholders or require reconsideration of the options available for management. For this reason, the decision criteria should include consideration of data adequacy, i.e., how much uncertainty in the risk estimate can be tolerated? How much data will be needed to support a sound decision or risk management option?

Data compilation and analysis

The element of "data compilation and analysis" is the lynchpin between the risk management and risk-assessment process (Chapter 1; USEPA 1992, 1998). Environmental risk assessment is described as an iterative process in which data needs are successively identified and implemented to better inform or refine the assessment. Here we refer explicitly to the technical data generated through the ERA process, but we recognize that analytic processes may be equally important to evaluate the economic, legal, and technological dimensions of the decision. Suffice it to say that a "gap analysis" may aid in identifying additional information needs (Pittinger et al. 1998).

The degree to which effective interaction occurs between risk managers and risk assessors in the earlier elements of the framework process ultimately determines the relevance and utility of the risk assessment in decision-making. Information must flow in both directions between the risk managers and the risk assessors. The risk manager must communicate to the assessor the perspective and the input gained from stakeholders through completion of the first three elements of the framework. Too often this is ignored since the technical personnel (often contracted services) used to conduct the assessment may be latecomers to the issue and are unfamiliar with the concerns of local stakeholders. The result may be an assessment conducted with little relevance to the decision at hand, the management options on the table, or the decision criteria to select among those options. For this reason, it is important that the assessors know not only what they are expected to do, but also why they are expected to do it, i.e., how the data will be used.

The risk assessor in turn has communication responsibilities to the risk manager. In advance of the assessment, the manager must understand clearly the scope of the assessment that reasonably can be expected on the basis of the resources and timing at hand. Managers must understand that few, if any, ERAs illuminate a single "right" decision or option. Typically, they provide a means to sort the data rationally in ways that distinguish causal mechanisms and ecological consequences of action or inaction. The assessor should be able to communicate how the assessment endpoints will relate to the management options that are identified and to the decision criteria that will be applied. The risk manager must further understand what the limitations of the assessment will be, e.g., making assumptions around exposure and effects data, limits to extrapolation to other species or media, natural variability, and uncertainties in the risk estimate.

How does a risk management decision affect the scope of the risk assessment? Clearly, the effort and expense of the data-collection process should be reasonably commensurate with the decision at hand. These efforts are often dictated by needs for accuracy or predictive certainty. Iteration may occur as data are collected, options are characterized, and the need for additional data are identified. While some iteration is appropriate and necessary, at times the iteration process between risk assessors and risk managers can become overly exaggerated (Pittinger et al. 1998). When perceived as circular or redundant, stakeholders may feel disenfranchised with the process and lose interest.

There is also concern in many cases that ERA serves as a "delay tactic" by risk managers who either fail to comprehend the decision that must be made or are unwilling (or incapable) of making such a decision. For this reason, the depth of study, the technical resources allocated to it, and the duration of the analysis should be commensurate with the priority and impact of the decision at hand. These issues should be recognized and discussed at the start, especially in management decisions involving multiple stakeholders, to promote the understanding that a finite decision, or action, hinges upon a given outcome in the analysis.

Once the assessment has been completed, it is incumbent upon the risk assessor to communicate the findings in a transparent and understandable manner. The extent of detail communicated regarding the methods, models, and statistics depends, to a large extent, upon the technical expertise of the risk manager. In addition to communicating results and conclusions, risk managers often require an understanding of alternative or additional investigations that may supplement the risk assessment and of the new information or insight these investigations might bring. Data gaps can be prioritized and addressed in an iterative manner, recognizing that complete knowledge is virtually unattainable and that some risks are inevitably unknown or non-quantifiable. It is usually the risk manager's responsibility to, in turn, communicate the findings of the assessment to stakeholders, but sometimes the risk assessor may become involved.

Option selection ("decision-making")

Option selection is synonymous with decision-making, the ultimate goal of the ERM process. The risk managers must consider and weigh the input of stakeholder groups and drivers to the decision in selecting the "most reasonable and acceptable option" under a given set of circumstances. The context of the risk, the attributes of the natural resource, and the stakeholders' perception of the resource's value largely determine what may be deemed acceptable or unacceptable. It is here that the risk manager will apply the set of weighted criteria derived in the third element, "management options development," using them in combination with the information derived through data-compilation and analysis to make an informed decision based on socially derived criteria of acceptability.

The "reasonableness" of an ERM decision (i.e., its acceptability or unacceptability) is viewed as a function of the social (perceived) value of the ecological entity or natural resource at risk. The resource's perceived value ideally is reached through a collective consensus among various stakeholders groups. Other factors, such as the severity and irreversibility of the risk and the stringency, cost, and equability of the proposed management option, also play important roles in determining acceptability (Cooper 1998). A conceptual example of such a relationship (Figure 2-2) relates

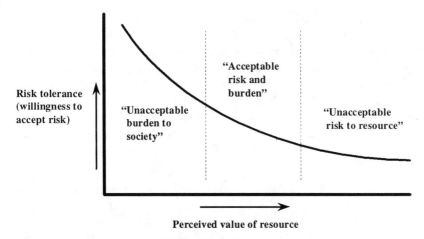

Figure 2-2 Relationship between risk tolerance and the perceived value (e.g., in terms of function, scarcity, uniqueness, inaccessibility, etc.) of an ecological resource

the value given to an ecological resource to the acceptability of a risk management option, i.e., risk tolerance. In this relationship, a resource's value may be expressed by a variety of scales. For example, an attribute of ecological structure or function may reflect its importance in maintaining ecological integrity (e.g., as a primary producer). Other measures of resource value may reflect more closely economic values related to parameters such as scarcity, uniqueness, vulnerability, or inaccessibility (see Chapter 6). Consistent with the law of supply and demand, an ecological entity or natural resource that is considered relatively resilient to harm, in abundant supply, or easily accessed tends to be less valued (either by ecologists or by the public) than those perceived to be relatively more vulnerable, less abundant, or inaccessible. Other measures or resource value may revolve around the cultural value of the resource (e.g., bald eagles).

The conceptual relationship depicted in Figure 2-2 suggests that risk tolerance, the acceptability of subjecting a natural resource to risk, is inversely related to the resource's perceived ecological or social value. That is, decisions to undertake or implement more stringent, restrictive or expensive risk management measures will receive greater acceptance for highly valued resources than for less valued resources. Ideally, the risk manager should strive to identify and select management options that represent "acceptable and reasonable risk," i.e., those that would be viewed as neither underprotective of the resource nor overly burdensome to stakeholders. More often than not, however, important risk reduction or avoidance decisions are difficult, and disagreements among stakeholders are inevitable. Clearly, this relates to the points raised earlier regarding tradeoffs among quality-of-life dimensions, as well as the importance of a considered and well-managed stakeholder-participation process.

The role of the risk manager and stakeholder may differ in "option selection" relative to other elements of the management framework (Pittinger et al. 1998). Many at the SETAC workshop, who helped to develop the risk management framework, agreed that the authority and accountability for a management decision must be retained by those with ultimate responsibility for the decision. While risk managers should carefully weigh stakeholder input, at some point a decision must be made by the risk managers without undue influence by a single stakeholder group. Examples were cited in which controversy arising from continued stakeholder advocacy, even in the final option selection, was seen as a barrier to effective and definitive decision-making.

At the same time, inability or failure to make a definitive or timely risk management decision may result in ecological and economic consequences that are as serious as a poorly informed or hasty decision. Concerns were raised that some regulatory managers seemed to seek highly compromised solutions that reflected an impractical marriage of all options, leading to ineffective or unsatisfactory management. The need for practical training in decision-analysis techniques and risk management is very important.

Decision implementation

Implementation of the management decision sets in motion the regulatory, legal, and economic processes required by the option selected. It requires that the risk managers have adequate authority not only to make the decision, but also to enforce or delegate enforcement authority for the decision to other parties. A tentative implementation plan for each option should be developed, documented, and made available to the interested publics prior to the decision (i.e., during the options development phase) to ensure that stakeholders are reasonably aware of what each option will entail. Problems in implementing decisions are often economic in nature; for example, many Superfund site remediation decisions have incurred costs that have far exceeded original expectations. Other difficulties encountered have arisen from management plans that fail to deliver the degree of resource protection originally envisioned.

Tracking and evaluation

The uncertainty and variability in ERA and ERM set limits to predicting the effectiveness of management decisions, often triggering the need for follow-up evaluation after a decision is implemented. Evaluations will range in complexity from simple periodic observations to sophisticated ecological monitoring programs. The criteria against which the effectiveness of the risk management activity is evaluated (the evaluative criteria) should be developed in the context of the criteria used to select the management option. That is, both the selection and the evaluation process should be based upon notions of "acceptable risk" with all its overtones of scientific method and of social valuation. The evaluation thus should consider such dimensions as the severity and extent of the risk, the reasonableness of the measures taken for evaluation, and the acceptable level of uncertainty. Finally, some mechanism should be constructed to close the loop between the evaluative process and the selection of options to ensure that recommendations are incorporated into subsequent risk management decisions.

The framework for ERM, briefly described above and by Pittinger et al. (1998), was the collective effort of the SETAC workshop held in Williamsburg, Virginia, in 1997. It was envisioned that such a framework would describe the fundamental elements of the process through which public and private risk managers determine what natural resources are to be protected from a given risk and the degree to which those resources should be protected. Individual elements and aspects of the framework are discussed in detail in the following chapters and illustrated through an informative series of case studies.

References

[ASTM] American Society for Testing and Materials. 1997. Proposed guideline for ecological risk management. Conshohocken PA: ASTM.

Cooper WE. 1998. Risk assessment and risk management: An essential integration. *Human Ecol Risk Assess* 4:931-937.

Engi D, Glicken JX. 1995. The vital issues process: Strategic planning for a changing world. Albuquerque NM: Sandia National Laboratories. SAND95-0845.

[NRC] National Research Council. 1983. Risk assessment in the federal government: Managing the process. Washington DC: National Academy Press.

[NRC] National Research Council. 1996. Understanding risk: Informing decisions in a democratic society. Washington DC: National Academy Press.

Pittinger CA, Bachman R, Barton AL, Clark JR, deFur PL, Ells SJ, Slimak MW, Stahl RG, Wentsel RS. 1998. A multi-stakeholder framework for ecological risk management: summary of a SETAC technical workshop. Summary of the SETAC Workshop on Framework for Ecological Risk Management; 23–25 June 1997; Williamsburg, VA. Pensacola FL: SETAC.

Power M, McCarty, LS. 1998. A comparative analysis of environmental risk assessment/risk management frameworks. *Environ Sci Tech* p 224–231A.

Presidential/Congressional Commission on Risk Assessment and Risk Management. 1997. Framework for environmental health risk management, Final Report, Volume 1. Washington DC.

Saaty TL. 1990. The analytic hierarchy process. Pittsburgh PA: RWS Publications.

[USEPA] U.S. Environmental Protection Agency. 1992. Framework for ecological risk assessment. Washington DC: EPA-630-R-92-001.

[USEPA] U.S. Environmental Protection Agency. 1998. Guidelines for ecological risk assessment. Washington DC: EPA-630-R-95-001F.

CHAPTER 3

Stakeholders and the Ecological Risk Management Process

Jessica Glicken Turnley

An underlying element of the ecological risk management (ERM) framework described in Chapter 2 (Figure 2-1) is the concept that stakeholder communication occurs throughout the risk management process. Effective management of stakeholder participation will depend upon the level of appreciation risk managers have for the role of stakeholders in the ERM process itself and their corresponding willingness to devote appropriate resources to stakeholder inclusion and communication activities.

It should be emphasized early on that stakeholder participation in decision-making does not relieve risk managers from their role as decision-makers. Rather, stakeholders are contributors of information to the decision-making process; generally, this information is about community values as they relate to ecosystems and associated resources. In general, information provided by stakeholders does not substitute for the scientific information provided by an ecological risk assessment (ERA) or other means of collecting scientific data, although there is some concern among environmental and business groups that the government's increased use of stakeholder-based decision-making is a "retreat" from more scientifically based rigorous methods (Yosie and Herbst 1998).

Stakeholders and Acceptable Risk

The core of the ERM process is the identification of management options that will lead to some level of "acceptable" risk. As noted earlier, science (i.e., the risk assessment portion of the ERM process) can determine the extent of adversity through its characterization of that risk. The public (i.e., all stakeholders) determines the acceptability of risk. This acceptability is defined through a series of tradeoffs made by society of the risk's social, economic, cultural, political, legal, and ecological costs (a discussion of ecological cost and benefit is provided in Chapter 6).

Acceptability depends heavily on assessments of how much an action that avoids or mitigates risk is "worth" to society as expressed through definitions of quality of life. "Acceptable" as it is used here in the context of "acceptable risk" becomes a relative, not an absolute, term because it is time- and space-specific. It also is dependent upon a definition of quality of life with its accompanying metrics. What is the social cost of the reduction or elimination of the risk? What is the social cost of allowing the risk to remain or escalate? By how much is the public willing to reduce its economic well-being to retain biodiversity? How important is it to invest social resources in fighting wars (maintaining our political security) if they result in long-term damage to the environment?

Environmental quality, economic well-being, political security or safety, and human health are generally recognized as elements of quality of life. Much of the disagreement in the debates on quality of life (Rosenberg 1995; Rogerson et al. 1989) center not on the elements that compose it but on metrics for valuing or defining those quality-of-life elements. The ERM framework developed at this workshop suggests that while the value of the metrics may be time- and space-specific (what level of economic well-being is "enough"? How safe is "safe"? How clean is "clean"?), the process for determining these values is not.

Quality of life can be understood as an integration of conditions of the world, including ecosystem integrity, with the experience of those conditions by individuals (Glicken 1996). A discussion of "external conditions" would depend heavily on the performance of ERA and other types of scientific studies and social analyses. This discussion is positivist and empirical in nature. The experience of those conditions by individuals assesses their perceptions of welfare or well-being, and is often described as a "welfarist" approach (Ringen 1995). This approach depends heavily on self-report, i.e., on elicitation of reports of personal experiences, attitudes, or activities, as distinguished from the reports by observers found in positivist approaches.

The welfarist dimension of the quality-of-life concept emphasizes the importance of stakeholder involvement in environmental decision-making processes. Stakeholder input is critical to identification and definition of the criteria by which these tradeoffs will be made (the "metrics" and their values for quality-of-life dimensions) and the relative weights that need to be given to the criteria. This input should be objectively considered by the risk manager in order to decide which option, or combination of options, offers the best single solution to all collective criteria.

Identifying Stakeholders

A poorly managed inclusionary and communication process can lead to inadequate or inappropriate stakeholder selection or information elicitation. The identification of those with whom the risk manager communicates and the development and implementation of communication vehicles all can impact the effectiveness of the risk management process. Yosie and Herbst (1998) note that the future uses of stakeholder processes will face five significant challenges:

- achieving quality management of stakeholder processes,

- measuring stakeholder processes and results,
- engaging the scientific community in stakeholder processes,
- integrating stakeholder deliberations with existing decision-making processes, and
- determining whether stakeholder processes will yield improved decisions.

All of these challenges relate to the rigor as well as the quality of the process that is implemented.

Identifying which stakeholders are appropriate to include in an ERM process can itself be a challenge. There are many ways to define stakeholders. Babiuch and Farhar (1994) give 16 definitions of 'stakeholder' that speak of "groups," "parties-at-interest," "people," and other similar terms and reference many more. For purposes of this discussion, we will use the following definition, which captures many of the concepts addressed in Babiuch and Farhar (1994). This definition has been well-received by a wide variety of groups when field tested: a stakeholder is an individual or group influenced by—and with an ability to significantly impact (either directly or indirectly)—the topical area of interest (Engi and Glicken 1995).

One common feature of all definitions of stakeholders or interested parties is that they identify and define groups relative to a specific issue. That issue may be as broad as "the environment'" or as narrow as "the riverine ecosystem between A and B streets in Anytown." While some of these groups may exist over time as formal organizations (i.e., the Sierra Club or Chamber of Commerce), they become stakeholders only in reference to a particular issue. "Stakeholder," therefore, is a relative term. Disputes are resolved situationally by dialogue among the interested parties, not through reference to an external, abstract code such as a legal system (other than that which sets the context for the debate); therefore, the outcome of a particular public participation process is not precedent-setting for other instances of process implementation. Instead, it simply generates an agreement for particular people in particular circumstances at a particular time —an agreement that will change as the "people" (i.e., the interested parties), local values, and, therefore, issues change, as they will over time (Bear 1994). This site- and issue-specific nature of participatory processes has led to a wealth of case studies in the literature (Beatty 1991; Hayton 1993; Renn et al. 1993; Sample 1993; Kangas 1994; Finney and Polk 1995;) as well as theoretical treatments of the process itself (Brenneis and M'Gonigle 1992; Berry 1993; Davis and Wurth 1993; Bear 1994; Dale and Lane 1994; Webler et al. 1995). Therefore, those managing a public-participation activity will need to draw from a general suite of methodologies and tools that can be combined and then applied in a situation-specific manner.

Because stakeholders are defined relative to an issue, a clear definition of the issue is a critical first step. The understanding of the definition of that issue by a community will determine the community's acceptance of the appropriateness or inappropriateness of certain groups as stakeholders. For example, much of the contention around the opening of the Waste Isolation Pilot Plant (WIPP[1]) in southern New Mexico has putatively revolved around environmental issues related to potential seepage of radiation into ground water. However, many of the stakeholder groups see the issue as one of nuclear disarmament and environmental concerns a rhetoric vehicle to describe the general destruction caused by nuclear weapons and their design, manufacture, and storage. For stakeholder groups defined by antinuclear sentiment, the full stakeholder community and its resultant dialogue looks quite different than the stakeholder community delineated by stakeholder groups focused on strictly environmental concerns such as the potential seepage from a particular site.

The description of the social universe known as "stakeholders" to a particular issue can be constructed through iterative interviews with various communities. Communities are queried as to the type of groups they see as relevant and their relationships to each other. The groups are then grouped with others like them into social categories such as "user" or "regulator" and mapped in a format similar to that shown in Figure 3-1.

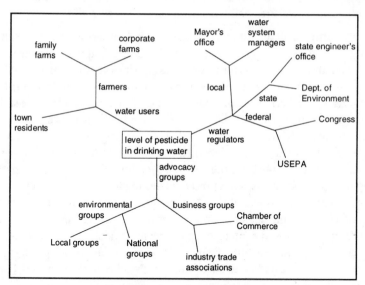

Figure 3-1   Example stakeholder map

The stakeholder map, in Figure 3-1, shows both social categories relevant to a particular issue and the relationship of those categories to each other. The smallest "fingers" of the map can be used to identify particular individuals if necessary. This stakeholder map can be modified or annotated based on the level of interest or political will various groups may have in the issue at hand. The steps in the process

---

[1] WIPP is the storage site for mixed radioactive waste from the national nuclear weapons complex.

for constructing such a map and identifying stakeholder groups are illustrated in Figure 3-2 .

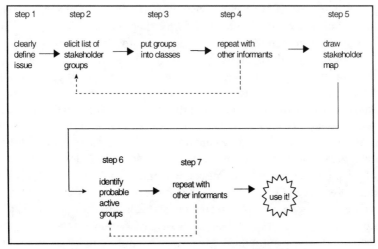

Figure 3-2   Process for constructing a stakeholder map

Communicating with Stakeholders

Once stakeholders have been identified, the risk manager must develop and implement a managed communication process. It is important from the outset to recognize the difference between a communication event and a communication process, the latter of which is composed of a series of events. The announcement of a public hearing, for example, is an event. The hearing itself is another event. The two together are part of a process to engage the public and should be managed along both dimensions. Failure to distinguish the event from the process generally leads to disappointing outcomes and does not allow the risk manager to correctly and appropriately evaluate the effectiveness of individual communication events. To follow on the earlier example, did the participatory process become contentious because some stakeholder groups did not find out about the hearing and therefore did not attend? Or was the hearing itself poorly moderated so that not all voices were given an opportunity to be heard? In the first case, it was the notification event that failed, and in the second, the hearing itself. In each case, corrective measures would be different. The ERM framework, as shown in Chapter 2, proposes a communication process that goes on throughout the risk management decision process.

Other common mistakes in stakeholder communication processes arise from a lack of knowledge about the audience or the target of the communication and from the application of "mirror imaging," i.e., assuming that "they are like me." This often leads to encoding mistakes, i.e., the use of vocabulary inappropriate to or not understood by the audience. This is particularly important when managing the interface between scientific or technical communities and lay audiences. Scientists and other technical individuals often resist what they believe to be the "dumbing down" of their information through its translation into a lay vocabulary. Lay audiences may perceive the use of such vocabulary to be a deliberate attempt at obfuscation or, at best, feel that they did not receive the information they needed to allay any concerns they had or to answer any questions. Conversely, scientists in a risk management situation often discount the value-based information provided by a community in favor of technical data. The community that decided to continue to allow hunting of an animal whose population had reached dangerously low levels in a particular area according to scientists is an example of a decision reached through a social tradeoff whose outcome seems counterintuitive or "wrong" to the scientific community. Renn et al. (1991) shows us that as the participants' perceived stake in the outcome of the issue becomes greater, nonscientific knowledge (values) plays an increasing role in the debate and the relative weight of science grows corresponding-ingly smaller. Effective communication between scientific and lay communities is critical for an acceptable and accepted risk management decision.

An ERM decision comes at the end of a long and complex process (chapter 1). If nontechnical stakeholders are treated as ancillary to the process by the risk man-ager, discounted in favor of scientific contributors, they are much less likely to accept the decision when it is made. If they are included in the entire process through substantive contribution where appropriate (such as in the determination of assessment endpoints) and by periodic communications as to the progress of the process, they will be much less likely to dispute the decision made.

Other chapters in this book address key points of stakeholder inclusion, particularly during the ERM process. These occur early in the risk management process and are focused on the identification of the resources to protect and the setting of the objectives for the risk management decision.

References

Baubich WM, Farhar BC. 1994. Stakeholder analysis methodologies resource book. Golden CO: National Renewable Energy Laboratory [NREL]. NREL/TP461-5857.

Bear D, editor. 1994. Public participation in environmental decision making. Washington DC: American Bar Association, standing Committee on Environmental Law.

Beatty KM. 1991. Public opinion data for environmental decision making: The case of Colorado Springs. *Environ Impact Assess Rev* 11:29-51.

Berry JM. 1993. Citizen groups and the changing nature of interest group politics in America. *Ann Am Acad Prac Soc Sci* 528:30-41.

Brenneis K, M'Gonigle M. 1992. Public participation: Components of the process. *Environment* 21(3)5-11.

Dale AP, Lane MB. 1994. Strategy perspectives analysis: A procedure for participatory and political social impact assessment. *Soc Nat Resour* 7:253-267.

Davis FL, Wurth Jr. AH. 1993. American interest group research: Sorting out internal and external perspectives. *Polit Stud* (XLI): 435-542.

Engi D, Glicken J. 1995. The vital issue process; strategic planning for a changing world. Albuquerque NM: Sandia National Laboratories [SNL]. SAND95-0845.

Finney C, Polk RE. 1995. Developing stakeholder understanding, technical capability, and responsibility: The New Bedford harbor superfund dorum. *Environ Impact Assess Rev* 15:517-541.

Glicken J. 1996. Methodologies for defining "quality of life." Joint meeting of the European Association for The Study of Science & Technology and The Society for Social Studies of Science; 10-13 Oct 1996; Bielefeld, Germany.

Hayton RD. 1993. The matter of public participation. *Nat Resour J* 33:295-281.

Kangas J. 1994. An approach to public participation in strategic forest management planning. *For Ecol Manage* 70:75-88.

Renn O, Webler T, Johnson B. 1991. Public participation in hazard management: The use of citizen panels in the U.S. *Risk-Issue Health Saf* 197:197-226

Renn O, Webler T, Rakel H, Dienel P, Johnson B. 1993. Public participant in decision making: A three-step procedure. *Policy Sci* 21:189-214.

Ringen S. 1995. Well-being, measurement and preferences. *Acta Sociol* 38:3-15.

Rogerson RJ, Findlay AM, Morns AS, Coombes MG. 1989. Indicators of quality of life: Some methodological issues. *Environ Panning* 21:1655-1666.

Rosenberg R. 1995. Health-related quality of life: Between naturalism and hermeneutics. *Soc Sci Med* 41(10):1411-1415.

Sample VA. 1993. A framework for public participation in natural resource decision making. *J For* 93:22-27.

Webler T, Kastenholz H, Renn O. 1995. Public participation in impact assessment: A social learning perspective. *Environ Impact Assess Rev* 15(5):443-463.

Yosi TF, Herbst TD. 1998. Using stakeholder process in environmental decision making: An evaluation of lessons learned, key issues, and future challenges. Washington DC: Rudder Finn.

CHAPTER 4

Deciding What Resources to Protect and Setting Objectives

Anne L. Barton

Introduction

This Chapter deals with the first three of the steps listed in Chapter 2: issue identification, goal-setting, and management options development (Table 4-1), with emphasis upon setting objectives and, especially, deciding which ecological resources to protect.

Together, these steps set the scene for the process as a whole. In particular, they define the risk and other questions to be answered. Much of the recent debate about the appropriate use of risk assessment in ecological policy analysis revolves around this issue of what question is to be answered. To be technically tractable, rigorous, and scientifically credible, the risk question is often delimited in fairly narrow, technical terms, usually by the risk assessors themselves. Such narrowing, however, may greatly diminish the relevance of the assessment to the fundamental policy issue. The process described in this chapter asks that the policy issue be defined first. At times, this may lead to assessment questions that are less scientifically tractable than the risk assessors would wish, but it does define the needs of the policy decision and sets out the issues that science must tackle if it is to be useful to risk management.

These steps provide guidance for the remaining steps in the process, but they do not determine the remaining steps nor prejudice the decisions. The objectives, for example, articulate what values are at stake; they do not limit the decision. The management options that appear to be available are listed to guide the analysis and alert stakeholders and the public. This does not preclude the development of additional options later in the process.

Table 4-1 Substeps for Steps 1 through 3

Issue identification	Goal-setting	Management options development
Initiating force	Problem statement	Gap analysis
Mandates	Overall goal	Draft options and criteria
Public involvement	Specific objectives (ecological/other)	Screened options
		Developed criteria

Risk Management: Ecological Risk-Based Decision-Making. Ralph G. Stahl, Jr. et al., editors.
©2001 Society of Environmental Toxicology and Chemistry (SETAC). ISBN 1-880611-26-0

Step 1: Identify the Issue

The three elements of issue identification, goal-setting, and management options development typically occur in unison in an ecological risk management (ERM) process. Issue identification is the traditional first step in decision-making, where the need for a decision is described.

The success of a risk management process depends heavily on the ability of all participants to understand the context, purpose, and limits of that process. Such understanding can be greatly improved by an initial identification and clear statement of the issue, including its initiating forces, mandates, and concerns of the various participants and interested public.

Initiating force

What has initiated this set of activities? Issues are generated through a number of mechanisms, including commercial enterprises and government programs (e.g., by Department of Defense, Department of Education, the U.S. Army Corps of Engineers), legislative requirements (e.g., the explicit protection of endangered species), public concerns raised in various forums, advocacy by interest groups, the dissemination of new information, or the availability of new technology. For example, in the DuPont Superfund project (Chapter 11) the initiating force was the listing of the site on the Superfund National Priorities List and the issuance of a U.S. Environmental Protection Agency (USEPA) Record of Decision (ROD) requiring DuPont to address risks of the wastes that had been land filled at the site.

How an issue may arise frequently determines its course of management and the degree of urgency given to its resolution. For example, negotiated issues that arise through the course of public debate often contain more options for resolution than do legally or administratively defined (regulatory) issues.

Mandates

The very first step in the ERM decision process is to articulate the initiating force and the legal (or other) mandates and restrictions associated with it. This includes an explicit identification of the decision-maker and a clear definition of the role stakeholders will play.

The USEPA ROD provided the mandates for the DuPont Superfund projects in Chapter 11. For example, the USEPA required DuPont to excavate contaminated sediments in a wetland area and restore the excavated area to grade using clean fill material.

Public involvement

As pointed out in Chapter 3, the interested public should be brought in at the very beginning of the process. In some cases, this may mean forming partnerships. In others it may mean communicating the issue to the public and soliciting views and information from them. In both cases, identification of the appropriate public is important. (See Chapter 3 and Glicken 1999 for a discussion of the stakeholder inclusion process.) During this step it is important to collect the information about "who cares" and "what do they want" needed for Step 3.

In the DuPont example, the DuPont remediation team met with the USEPA, state regulators, and other stakeholders to discuss goal-setting and how best to implement the requirements of the USEPA ROD.

Step 2: Setting Goals and Objectives

During goal-setting, the environmental issue is translated into a series of goals and objectives that are conducive to systematic analysis and planning by risk assessors, natural resource managers, economists, planners, etc.

Goal-setting must recognize not only ecological goals, but also economic, human-health, political, legal, and social objectives as well. The social necessity for tradeoffs between environmental quality and economic well being—as in the well-known case of the spotted owl—often is highlighted in ERM decision-making. In a more extended example, a rural farming community concerned about the protection of a rare riverine ecosystem might first develop a succinct statement of the issue which describes the conflicts and challenges between the economic goal of farming and the effects of farming upon the endangered ecosystem. A general goal, or ideal outcome, might be the maintenance of a successful farming economy and the simultaneous protection of the ecosystem. Specific objectives supporting this goal, e.g., maintaining irrigation water volume or habitat for an endangered species, can be linked to measures for the risk assessors. See the quality of life discussion in Chapter 3, where ERM decisions are treated as the outcome of a series of tradeoffs among potentially competing social objectives.

What are objectives?

Keeney (1992) defines "objective" as "a statement of something that one desires to achieve. Three features characterize it: a decision context, an object, and a direction of preference. For example, with respect to traveling in automobiles, one objective is to maximize safety. For this objective, the decision context is automobile travel, the object is safety, and more safety is preferred to less safety." For environmental and some other objectives, a desired state may take the place of the direction of preference (e.g., sufficient habitat to support the 1970 population level, rather than maximize habitat).

These concepts may appear with different terminology than that used here. For example, The Nature Conservancy's five S framework refers to the object of an overall goal for a site as the "conservation target" (see Chapter 13). The overall goal for a site is thus "to maintain healthy viable occurrences of the conservation targets and meet human community needs." Once the conservation targets have been identified, the specific objectives are articulated by characterizing the desired size, condition, and landscape context that represents a viable occurrence for each target.

Objectives hierarchies

Objectives hierarchies organize objectives in such a way that the lower level objectives are contained in and collectively define the higher level objective. One example (from Keeney 1992) is in the decision context of setting a carbon monoxide air quality standard. If the overall objective is to minimize carbon monoxide concentrations, objectives at the next lower level could be to minimize carbon monoxide concentrations in industrial areas, to minimize carbon monoxide concentrations in urban areas, and to minimize carbon monoxide concentrations in rural areas.

Objects, instead of fully stated objectives, can be organized into hierarchies with fewer words. Keeney gives an example in the decision context of transporting nuclear waste. Here the overall object is health and safety. Under this, there are two lower level objectives: future generations and current generation. The value for current generation is further subdivided into cancer and noncancer. Finally, each of these is divided into public and worker.

Some benefits of organizing objectives or values into hierarchies include ensuring that nothing essential is omitted, clarifying and confirming the meaning of the higher level objectives or values, and showing the relationships among objectives. Objective hierarchies can also be constructed in order to develop a quantitative value model.

In this chapter, we suggest that the overall goal be made more specific in lower level hierarchies. In the previously mentioned farming community, the overall goal of maintaining a successful farming economy and protecting the ecosystem is specified in objectives such as maintaining irrigation water volume or habitat for an endangered species. However, the process can be repeated as many times as necessary until it reaches the level of specificity that is most useful for assessment and decision- making.

In the section on deciding what to protect later in this chapter, the first level of an object hierarchy is developed for the overall object the (natural) environment. At the first level, the natural environment is divided into: "plants, animals, and their habitats"; "whole ecosystems: their functions and services" and "special places." Each of these categories can be further subdivided at least once. For example, "plants, animals, and their habitats" can be divided into "populations," "communities," and perhaps other categories. "Whole ecosystems, their functions and services" can be divided into "ecosystem types" and "functions and services." Obviously, these could be further subdivided into specific types of ecosystems and specific functions and services.

Means–ends hierarchies

As defined above, objectives are statements of something that one desires to achieve. However, some objectives are fundamental ends, while others are means of accomplishing more fundamental objectives. For example, if your fundamental objective is to "maximize automobile safety," one means of doing this is to "minimize automobile accidents." This means objective can be further broken into means such as "minimize driving under the influence of alcohol." A set of objectives starting with fundamental objectives and ending with the means objectives most immediately related to a decision is a means–ends hierarchy.

Sometimes a project starts with objectives which appear to be means rather than ends (e.g., reduce emissions). To trace the end (fundamental) objectives for specific means objectives, you repeatedly ask the question Why is this important? One answer is that it is one of the essential reasons for interest in the situation. In this case, it is likely to be a fundamental objective. Another answer is that it is important because of its implications for some other objective. In the latter case, it is a means objective and the response identifies another objective (which itself may be either a fundamental objective or another means objective).

Keeney 1992 gives an example for a decision situation involving the transportation of hazardous material. One objective may be to minimize the distance the material is transported by trucks. The question that should be asked is why is this objective important? One response may be that shorter distances would reduce the chances of accidents. Then the question is why is it important to have fewer accidents? An answer could be to reduce the exposure of the public to hazardous material. (At this point, an issue may arise that shorter distances may not reduce exposure if they involve routing through major population centers. This might require some adjustments in the earlier means objectives.) The next question is why is it important to reduce exposure? The answer may be that it is to minimize the health impacts of the hazardous material. A response to the question why is it important to minimize health impacts may be it is simply important for its own sake. This indicates that you have probably found the fundamental objective.

The Nature Conservancy's five-S decision-making framework (see Chapter 13) employs means–ends hierarchies. The fundamental objective in this case is to "maintain healthy, viable occurrences of the conservation targets." The two general means employed to do this are "1) abate the critical sources of stress and 2) directly restore, enhance, or maintain the systems."

Substeps

Setting objectives starts with a clear statement of the problem, issue, or opportunity and ends with a set of specific objectives, that will guide all of the remaining steps. In this, step one answers the what to protect question for ecological resources and summarizes what is at stake with respect to all of the drivers. An example is given for a hypothetical community that is based upon a farm economy and also contains a rare riverine ecosystem. The substeps for this step are problem statement, overall goal, and specific objectives.

Problem statement

Make a clear, succinct statement of the problem arising from the issue based upon the information gathered in Step 1. In the example of the community mentioned in the previous paragraph, this would mean describing the conflicts and problems that have appeared in the community because of the effects of farming and the needs of farmers on the rare ecosystem, including any legal issues involved. This statement should describe the problem as the stakeholders perceive it. There is no need for analysis at this stage to determine the degree that various farming practices actually impact the ecosystem or what the alternatives are. This reality checking comes in step 4.

Overall goal

Set a general goal for the decision. This is a general statement of the desired outcome that would solve the problem or take maximum advantage of the opportunity. In the riverine community example, this might be to maintain both the traditional farm economy and the rare ecosystem in the area. The goal may or may not be obtainable. This is what is desired and serves as a guide for all the following steps. It is quite different from a final decision.

The overall goal for the DuPont Superfund example (Chapter 11) was to meet the legal requirements of USEPA's ROD by means that are safe, cost effective, and acceptable to the USEPA and the public. And to do so by means that protect human health and the environment and that are based upon sound science and information. The overall goal for the Waterfowl hunting example (see Chapter 8) was to maximize the sustainable harvest of the populations regulated.

Specific objectives

Specific objectives differ from the general goal because they should be sufficient to allow scientists to develop measures from them without distorting the objective. Objectives for the community described in Step 2 might include maintaining irrigation water volume sufficient for some given acreage of farms and maintaining habitat to support the 1970 population of an endangered species.

One of the objectives for the DuPont Superfund example was to provide an increased level of ecological services in the restored habitats by implementing a remedy that would exceed the requirements of the ROD at little or no additional cost.

The specific objectives of the waterfowl-hunting example (see Chapter 8) are included in the 6 goals of the regulatory process. They included providing an opportunity to harvest a portion of the game bird populations, maintaining the populations, and avoiding the taking of endangered or threatened species.

Purpose of objectives

Goal-setting requires risk assessors and managers to frame the issue in meaningful, measurable, and achievable terms to focus and accommodate specific data compilation objectives. Acknowledgment of the entity or resource to be protected, the attribute of resources essential to its protection, and the values represented by stakeholders should be communicated. Values of diverse interests may often be competing. For example, commercial waterfront property development may compete with the need to maintain an undisturbed wetland habitat to support aquatic communities. Competing interests and mutual beneficial interests should be identified and stakeholder representatives brought to the table early to maximize the acceptability of the analysts' results and the ultimate management decision.

Setting specific objectives helps to determine the scope and focus of the ecological risk assessment (ERA). This allows the assessment to provide information relevant to the risk manager's decisions and to do this in an efficient way (without the rework that often occurs when the risk manager concludes that a completed ERA fails to answer the necessary questions).

Setting objectives in advance of the assessment provides an opportunity for public involvement and the incorporation of public values at the beginning of the process. Having objectives from the beginning provides clarity to the risk manager, the risk assessor, and the public about the ecological and other values at stake in the decision and the information which the ERA is intended to provide.

Objectives do not prejudge the decision. Within the total set of objectives, some are likely to be in conflict with others because conflicting values are at stake for the decision. The objectives should reflect and clarify these conflicts, not attempt to cover them up or resolve them in advance of analysis and further deliberations.

If risk managers fail to set explicit objectives before the ERA begins, the risk assessors must set or assume them. This often leads to the use of implicit risk questions that can be answered technically, but are not relevant to the principle public issues.

Although many people and groups can be involved in setting objectives, they are risk management objectives and, thus, are ultimately the responsibility of the decision maker, which may be an individual, a government department, or an ad hoc group.

Types of objectives

Ecological objectives are not the only ones that need to be set in Step 2. Economic, human health, political, social, and other objectives should all be made explicit. These would be elicited through quality of life discussions as discussed in Chapter 3. By making all these objectives explicit, the process maximizes the usefulness of the various analyses and provides a basis for making and communicating trade-off decisions. Objectives for the example community mentioned in an earlier section include the maintenance of habitat to support the population of an endangered species as of some specific past year (ecological objective) and the simultaneous maintenance of irrigation water volume sufficient for some given acreage of farms (economic objective).

Ecological objectives

Some risk managers find the setting of ecological objectives particularly difficult because there is such a vast choice of ecological resources and so many different ways of looking at them. For example, many of the laws that govern the USEPA activities provide the general goal to protect human health and the environment without further definition of what is meant by "protect the environment." This section is intended to help risk managers focus on what to protect in such circumstances.

Ecological objectives are driven by the same factors as are other objectives and the decision process as a whole: legal, political, scientific, cultural, economic, and technical feasibility. Like other objectives, ecological objectives should include the object (what is to be protected or restored), the desired state (or direction of preference, such as "reduce" or "minimize"), and the decision context which connects the result to the decision being made. Such an objective is very useful to the ERA and allows risk assessors to select a set of assessment endpoints that will be useful in decision-making. Again, this is not the final decision and it may or may not be obtainable. An example of an ecological objective from Waquoit Bay (one of USEPA's Watershed Demonstration Projects) is reestablish a self-sustaining scallop population that can support a viable fishery. This goal may or may not be achievable. It may conflict with other goals for the area, such as development and tourism. However, it is a desired state and is useful in guiding the risk assessment and showing what tradeoffs are finally made. Other ecological objectives are provided in Table 4-2.

Factors that should be considered in setting ecological objectives are listed below.
- Resources of concern
  - plants, animals, and their habitats
  - whole ecosystems, their functions and services
  - special places
- Program or project mandates
- Precedents
- Stakeholder and public concerns
- Related ecosystem functions or components

Table 4-2 Examples of ecological objects and desired states

Category	Object	Desired state
Animals, plants, and their habitats	Aquatic communities in lakes, streams, and estuaries	Protect 95% of aquatic species from adverse effects on survival, development, and reproduction
	Habitat for a keystone species	Maintain habitat to support a sustainable population
	Regional populations of native species	Maintain a viable regional population
	Native and migratory birds or fish exposed to toxic pollutants	Avoid widespread and recurring or massive die-offs
	Scallops and their estuarine ecosystem	Reestablish a self-sustaining scallop population that can support a viable fishery
	Northern goshawk and its old-growth habitat	Sustain goshawk population
	Forest plants	Maintain plant abundance
Whole ecosystems and their functions and services	Flood and sedimentation control in stream corridors	Restore and maintain to a standard suitable for human activity in the area (e.g., farming)
	Wetlands	No net loss
	Soil fertility	Protect soil community to maintain soil fertility
	Estuaries productivity	Sustain production of economically important and food fish
	Biodiversity in the Everglades	Restore habitat mix to some optimum condition
	Lake Superior	Maintain as an oligotrophic lake
Special places	Remaining tall-grass prairie	Maintain current extent
	Water clarity of Lake Washington (in Seattle)	Restore and maintain water clarity to a specified past condition

Deciding What to Protect (Resources of Concern)

"The environment" is such a large concept and composed of so many interrelated things and processes that some risk managers have difficulty knowing where to begin. Furthermore, many of the ultimate protection decisions require consideration of the value society places on the resources under consideration.

Criteria and models for deciding what to protect

First, one must consider the range of ecological resources or values of concern that fall within the realm of the specific program or issue as described in the process step "issue identification." This could mean looking at the important resources within certain geographic boundaries (as in The Nature Conservancy example) or considering what ecosystems are likely to be exposed to and affected by the stressor being regulated (such as pesticides or other toxic substances).

Then, consider the mandates the program operates under. What does the law, previous regulations, or other prescribed actions require the program to protect? For example, all USEPA programs are covered by the Endangered Species Act.

The next criteria used is deciding what are the important precedents within the decision context. Do these suggest ecological values or objectives that must be included?

Next, what the ecological values are of particular importance to the stakeholders and the public must be considered. And, how do these overlap with the ecological values that fall within the range of the program or project needs to be answered.

Finally, the question of what are the ecosystems, ecosystem components, or ecosystem functions that have not been included in the answers to the previous questions but are important to the long-term status of what has been included must be answered.

In this section, we provide a set of three categories to help risk managers develop a list of potential entities for consideration (categories are from USEPA 1997). The categories overlap in the sense that any particular ecosystem, such as Lake Superior, might appear in all of them (examples are given at the end of this section). They are proposed not to classify the entities that are used, but as a checklist to guide the decision in the direction of what entities should be used. They are intended to help decision-makers decide what is wanted from an ecosystem (or to answer the "so what" question at the beginning).

Plants, animals, and their habitats

Some examples of this category are population of a valued specie; entire complex of plants and animals in a particular type of ecosystem, such as the "aquatic community" protected by the U.S. Water Quality Criteria; or full range of species associated with a locality. Sometimes groups of individuals are protected from death or deformity without specific consideration of the effect on populations or communities, such as when risk managers act to avoid large fish kills or presence of chemicals leading to deformed wildlife.

To protect species at the population or community level, one should consider the integrity of the ecosystems that support them, as well as direct threats to the plants or animals themselves. Otherwise, any gains made in protecting them from direct threat may be very short-lived.

Whole ecosystems: Their functions and services

This category views ecosystems as a whole. The associated goals are to maintain their geographic extent or special character or to preserve their value for providing certain services to humans or other ecosystems. The goals focus on the movement of energy and matter through the ecosystem in processes such as nutrient cycling or erosion control, rather than on specific components.

A well-known example in this category is wetland protection. The value of wetlands has been increasingly recognized over the past couple of decades. They provide some very important functions and services such as water-supply services, flood water regulation, shoreline anchoring and erosion control, water purification, and habitat for biodiversity.

Other ecosystems such as forests and grasslands also provide important services such as nutrient cycling, soil stabilization, and maintenance of air quality. Ultimately, the continued existence of all valued components (including humans) depends on the continuation of critical ecosystem functions. As the human population grows and greater portions of our natural resources are utilized, the value of these functions is likely to become more and more obvious to the general public.

Special places

This category includes specific geographic places of special interest because of their unique character or contribution to local or national natural resources. A lot of local ecological protection initially focuses on a special lake or other area which is important to the people who live, work, or visit in the community. It also includes areas such as endangered ecosystems, national parks and wildlife refuges, and ecological conservation areas, such as the Nature Conservancy's "Last Great Places."

The special character of the place needs to be carefully and explicitly defined so that it is clear just what is being protected. Many places are considered "special" because they provide habitat for rare species or for biological diversity. Others are valued for their aesthetic qualities or because they are among the few relatively pristine ecosystems of a particular type.

These categories should be reviewed to make sure all of the relevant entities are considered for protection. It is not important to assign any management objective to a particular category, it may be impossible to do so. For example, the Everglades can be seen as habitat for some valued species, as a particular type of wetland, and as a special place that can be located on a map. It is unnecessary to assign or restrict it to any one of these categories. Instead, review of all of the categories helps to define what is important about the Everglades. This information is useful in setting objectives. A good example of an objective that covers all three categories is the objective for Lake Superior in the U.S./Canadian Great Lakes initiative to "maintain Lake Superior as an oligotrophic lake with lake trout as the top predator" (International Joint Commission 1989). This mentions a valued animal (lake trout), the continuation of a particular type of ecosystem (an oligotrophic lake, one with high oxygen and low nutrient levels), and of course a particular geographic location (Lake Superior) that is important to citizens of both the U.S. and Canada.

Adding desired state or preferred direction

In order to set objectives, one needs not only the object (what to protect), but also the desired state or preferred direction. The decision of whether the objective is based upon a specific desired state or a more general preferred direction depends upon the values of stakeholders and decision-makers. For example, in the Waquoit Bay case, it may not have been sufficient to have a more general objective to maximize the scallop population. If the fishery is what is desired, then restoring the population to something less than what would support a fishery may be regarded as insufficient, but restoring it to any level greater than needed to support a fishery may be unnecessary. Science may also play a role in this (such as linking the scallop population in Waquoit to the eel grass abundance). Therefore, it is necessary to understand both scientific connections and social values early on.

The issue identification step (Step 1) may provide some information useful in deciding upon the desired state or preferred direction for specific systems, populations, or services.

1. What is the ecological value of the system, service, or population potentially at risk?

2. What is the public value of the system, service, or population potentially at risk?

3. Is this system or population unique in someway, and is this uniqueness a major determinant of the level of protection required?

Examples of Ecological Objectives

Table 4-1 gives examples of objects and desired states (or preferred direction). The examples are given according to the category in which the entity may have first been identified. Some of the examples could be associated with more that one category.

Step 3: Identify Risk Management Options

As noted in the initial framework discussion, this step asks the risk manager to develop an initial list of management options based on available (and possibly qualitative) knowledge and establishes a set of decision criteria by which the tradeoffs among various options will be made.

It is at this stage of the framework that the distinction discussed in Chapter 2 between data collection and interpretation (the ERA and other data collection activities) and the use of that data to make socially appropriate decisions (ERM) often begins to be confused. The ERA and various social analyses (described in step 4) should be used to describe the external conditions of the environment, using the goal-setting stage as a guide to the conditions on which to focus. The development of management options (treated here) and their subsequent selection (treated in step 5) will use the ERA information in conjunction with information of other types (such as economic, technical, legal, and others) to identify the preferred course of action.

Identify information needs (gap analysis)

Set out the available information gathered in Step 1 or previously known and use Table 4-3. If there is sufficient information to do a good job of identifying the options, then do so and list them. If not, gather the missing information and then list options. Although this gap analysis can serve as a guide for Step 4 (analysis), this step is not the place to do that analysis. Information needs to be sufficient only to list options, not to distinguish among them on merit or to make a final decision.

Table 4-3 Tabluar evaluation (gap analysis) of available information on a particular ecological risk management issue.

	Legal	Political	Cultural	Social	Science	Economic	Technical
What do we know?							
Who knows it?							
What do we still need to know?							
Who cares and what do they want?							

Draft options and criteria

Develop a list of options and outline decision criteria in conjunction with interested parties. This will be a fairly inclusive list of options which can be cut back in "screened options." Parallel to the development of a list of practicable management options, and equally dependent upon broad stakeholder input, is the establishment of clear decision criteria. The criteria should reflect the key concerns of stakeholders and the main drivers of the decision. A variety of decision analysis methods are available for identifying and weighting criteria in risk management decisions (see, for example, Engi and Glicken 1995).

Screened options

Select options for analysis based upon a qualitative screening for nonstarters (the "laugh test"). Various terminology is used to describe these screened options. For example, the Nature Conservancy's five-S Framework (Chapter 13) refers to the risk management options as a set of potential threat abatement and restoration strategies. The options available for adaptive regulations of waterfowl hunting (Chapter 8) have been organized into four general categories very restrictive, restrictive, moderate, and liberal.

Developed criteria

Decision criteria will depend, to a large extent, upon the context of the ecological risk and the affiliation of the risk manager, whether in a public (regulatory) or private (corporate) capacity. Both must satisfy multiple stakeholder groups. The regulatory risk manager must serve the expectations and concerns of the public, the regulated community, and elected legislative bodies, in a highly visible fashion often dictated by prescriptive statutes. Divergent concerns and sometimes inconsistent demands may surround the regulatory manager. The development of decision criteria provides an opportunity for making these explicit. Private risk managers are usually less subject to exhaustive public scrutiny, but also must answer to an array of stakeholders, including stockholders, regulatory bodies and the consuming public. Also, the development of decision criteria can make any conflicts explicit and reveal additional questions that must be answered by the risk assessment.

The process for indentifying risk management options may take the steps described here in a different order or combine some of them. For example, the DuPont Superfund example in Chapter 11 combined gap analysis and options screening in the matrix table described in the "management options development" section of that chapter.

Involving the Public in Early Stages

Public involvement is crucial to the setting of objectives and risk management options. This is a stage in which the involvement of social scientists could be very helpful (see Chapter 3 and Keeney 1992).

Influence of Setting and Application

Waste sites

One issue that appears in site management cases is what is or is not confined to the site itself. Generally, the risk management options are restricted to the site itself, but it is possible for the objects of protection to be outside the site.

There can be several risk managers for a particular site and all their interests are not likely to be the same. Place-based assessments are overall assessments with multiple stakeholders. Risk assessors need to come to the table often with risk managers and public to discuss what's going on at the site.

Natural resources

Resource managers generally know what they need to protect because their laws and organizations are devoted to specific environmental resources. Their laws may also specify the desired state or preferred direction in many cases. However, in other cases, the desired state requires additional deliberation. By contrast, the risk management options may be less obvious and may require the cooperation of partners or stakeholders.

Chemicals

Decisions about chemicals are stressor–based decisions. The toxicological receptors (individual plants or animals) become the focus of attention. This can work well in the case of human health where the individual organism is also the valued object (what to protect). However, in the ecological arena, the individual plant or animal is rarely the most valued object at stake. The real concern may be at a higher level of biological organization, such as population or ecosystems. By failing to go beyond the immediate toxicological endpoints, chemical decision-makers may unwittingly confuse the analysis and mislead themselves, their stakeholders, and the public about what is really at stake.

Therefore, it is particularly important in this setting to review all of the categories listed in "what to protect" and to consider not only individuals and populations, but also whole ecosystems and their services. These groupings need to be considered even when methods for estimating the risks to them are not immediately available. Scientists have a role to play in pointing out what ecosystem functions and services may be at stake in these decisions, but the ultimate decision of what to protect should be based on societal values.

Chemical decisions often have to be made within a very short time and therefore, the setting of objectives (and perhaps all of the first three steps) should be done on a generic basis. These steps would then form the background for testing and screening decisions that are normally made on a generic basis.

References

Engi D, Glicken J. 1995. The vital issues process: Strategic planning for a changing world. Albuquerque NM: [SNL] Sandia National Laboratories. SAND95-0845.

Glicken J. 1999. Effective public involvement in public decisions. *Sci Commun* 2:298-327.

International Joint Commission. 1989. Revised Great Lakes water quality agreement of 1978. Windsor Ontario, Canada: International Joint Commission.

Keeney RL. 1992. Value-focused thinking: A path to creative decisionmaking. Cambridge MA: Harvard University Press. 416 p.

[USEPA] U.S. Environmental Protection Agency. 1997. Priorities for ecological protection: An initial list and discussion document for EPA. Washington DC: USEPA Office of Research and Development. EPA-600-S-97-002. 44 p.

Selecting, Implementing, and Tracking Ecological Risk Management Decisions: Necessary Elements of an Effective Decision-Making Framework.

Peter L. deFur and David P. Clarke

This chapter details the four remaining steps in the ecological risk management (ERM) framework (Chapter 2): analysis, option selection, decision implementation and tracking, and evaluation. In addition, the elements necessary for an effective risk-based decisional framework are discussed along with some examples of how ERM decisions have been made in the past.

To be effective, the workshop participants agreed that risk-based decisions should take into account the full range and diversity of quantitative and qualitative information on social, legal, economic, and scientific aspects of the decision. Each of these areas is a discipline itself and has literature, data conventions, and methods for analyzing those data. Each area could easily be the subject of a treatise. Here we seek to express the level of integration and summarization necessary among the disciplines that must go into the ultimate decision-making for ERM situations. In particular, this chapter examines different features of risk-based decision-making as applied to the context of ecological decisions. These features include the scientific and technological issues, social, legal (or regulatory), and economic information.

A number of existing frameworks provide adequate guidance for understanding the elements of risk management and the associated process that must be known to intelligently manage ecological risk problems (i.e., Presidential/Congressional Commission 1997; USEPA 1995). Also, they identify the array of scientific and nonscientific factors that weigh in such decisions, but, they fail to provide much insight into three fundamental issues:

- What ecological resources should be protected?
- How much protection should each resource receive?
- How should the desired level of protection be achieved?

Instead of articulating a framework for "decision-making" on these three issues, existing approaches provide a framework for recognizing how infinitely complex these decisions can be. More importantly, most of these existing frameworks are designed to help manage risks to human health and, therefore, may not be fully applicable to ERM.

In addition to discussing the general elements of sound ERM, this chapter sets forth a framework for decision-making relying on the ecological sciences to provide the scientific inputs (see Chapter 1, Figure 1-1 for details). In addition, Chapter 6 illustrates how elementary economic principles provide much-needed structure to the ERM framework.

The framework for decision-making presented here is not a separate or distinct framework from the one developed by this workshop (Chapter 2), However, it recognizes that other factors besides ecological science inevitably will enter into decision-making and, for that reason, is provided as an example that may stimulate additional discussion and research in ERM. As discussed in Chapter 6, economic reasoning serves the critical purpose of forcing scientists to articulate tradeoffs among competing ecological risk objectives, establishing a conceptual device for specifying the rates at which such tradeoffs can be made, and identifying the "best" mix of ecological resources to protect for any given level of commitment to protection (e.g., setting a protection level at 95% of a population). More importantly, if further efforts will offer greater marginal benefits than marginal costs , economics will provide the tools for determining whether an existing case can continue this commitment to a specific level of protection.

General Elements of Risk-Based Decisions

As noted in Chapter 2 (and previously in Pittinger et al. 1998), the workshop participants reached a level of agreement on key elements of a framework for ERM. Several of these elements listed below, apply specifically to the decision mode and are indicated here as option selection (step 5) of the ERM. These elements were that

- the entire process needs to be focused on making a decision rather than analyzing data and gathering information;
- decisions should take a variety of types of information into account;
- the difficulty of making decisions increases with the magnitude of consequences;
- decisions that are small with respect to predetermined criteria (costs, scale, precedent) require less time and resources devoted to analysis making decisions has consequences that should be considered and that requires follow-through;
- these are implementing, communicating, and monitoring the results; and
- criteria and options for decision-making need to be sufficiently adaptable to specific cases and conditions, rather than overly rigid and prescriptive.

Several sources of information on risk-based decisions continue to provide new analyses and approaches, some even provide novel or innovative methods. The authors of this chapter relied on the discussions among participants in the workshop, the reports from the National Academy of Sciences/National Research

Council, the journal *Human and Ecological Risk Assessment*, and other published literature on the subject.

Defining risk-based decision-making

During the workshop, and in the preparation of each work group summary, the participants discussed ERM with the assumption that the members of the group all understood what "management" and "decision-making" meant. The preliminary materials (Stahl et al. 1997) and summary monograph (Pittinger et al. 1998) referred to ERM as the process by which individuals and groups use information to determine what resources to protect, to what degree, and how to provide that protection (see Chapter 4 for more details). When the discussion at the workshop turned to the decision-making element, the work groups and plenary discussion considered ways in which information is used to make decisions. With that discussion as the starting point, we will use the term "ecological risk-based decisions" to mean making decisions on the basis of the range of information collected in the process of assessing ecological risk. This definition is less than completely satisfying because it is still rather vague. Several categories of information are included in risk assessments and subsequently used in the management decisions. Information on exposure, effects, dose-response, and uncertainty should come from the assessment. Information on economics, socials issues, and legal requirements are not part of the technical assessment of risk, but go into the management analysis and decision. This chapter considers what is meant by risk-based decision-making, some of the elements of such decisions, and some of the processes used in making these decisions.

Generally, experts from various fields believe that the data in their field of expertise (e.g., ecology, toxicology, law, economics, and sociology) will certainly play a part in a risk-based decision; the workshop participants agreed on this point. Furthermore, this conclusion is hardly new and is similar to the approaches taken in at least two other Society of Environmental Toxicology and Chemistry (SETAC) workshops (Barnthouse et al. 1998; Reinert et al. 1998) that focused on sustainable environmental management and ecological risk decision support systems, respectively. In both of the previous reports, it is clear that experts believe a wide range of information must be used in making appropriate decisions in the context of ecological risk assessment (ERA). In the present workshop, the participants agreed on a Framework that indicates when and how diverse information should be used in ERM.

The process of risk-based decision-making

The participants in this workshop referred to a wide array of methods used to process information in making decisions in an ecological risk context. Some of the methods identified were

- comparing data,

- evaluating information independently,
- ranking and scoring data,
- filling in check lists,
- asking questions,
- examining completeness, and
- weighing various factors.

The specific process of making decisions based on ecological risk has many different meanings and incorporates a great variety of approaches and processes in deciding. The National Research Council (NRC), in the report Understanding Risk (NRC 1996a), identified two general categories of processes for such decision-making: analysis and deliberation. The National Research Council went on to recommend that the way to approach decisions in a risk setting is a combination of these two processes and used the term analytical-deliberative process in referring to the interplay between the two types of processes. In the context of risk-based decisions, the NRC report used deliberation as "any formal or informal process for communicating and for raising and collectively considering issues." Analysis was taken by NRC to mean "systematically applying theories and methods that have been developed within communities of expertise such as those of the natural science, social science, engineering, decision science, logic, mathematics and law." It was clear from the discussion that many participants at the workshop shared a similar perspective.

Certainly a substantial body of information has been developed in the field of decision science to elaborate the theories surrounding decision-making. We do not pretend to invent an entirely new set of theories and conditions separate from that body of knowledge. Rather, the goal of the workshop and of this publication is to add to that knowledge by describing elements of decision-making in ERM.

A major area of ERA and ERM in the U.S. has evolved within the federal family of agencies charged with implementing various environmental statutes as well as developing regulations and guidelines. Existing government and industry programs, notably U.S. Environmental Protection Agency (USEPA) program offices such as the Offices of Water, Solid Waste and Emergency Response, and Pesticide Programs, have a sufficient history and experience to describe their current practices for making risk-based decisions. Although many major environmental statues and regulations are placed under one agency, the development of these regulations and their implementation under the USEPA are program-specific (Superfund, Clean Water Act, Clean Air Act). As a result, the programs often operate under specific federal laws that prescribe some aspects of the analyses the Agency is required to conduct. In one program the analyses required may include a cost-benefit test that must be applied to a particular risk management action to determine whether the proposed action is cost effective. Yet, the same requirement may not be required under a different program. Having this existing "framework" imposed upon the

Agency can stifle flexibility in some cases. In addition, the law often sets out the criteria and procedures for how the Agency weighs information and reaches the final decision. Since laws seldom envision all situations where they may be applied, prompt responses to changing societal norms and environmental conditions can be stifled. Because of the difference in focus specified for the various USEPA offices, it is not unusual for there to be a wide diversity in approaches to ERM (USEPA 1994), which can lead some to voice concerns about the inconsistency that arises in decision-making (Slovic 1993).

Even though this diversity exists within the USEPA, the general practice is to conduct or require a permittee or applicant to conduct an ERA according to present guidelines and/or requirements (see Chapter 7 for more detail). Once the assessment is finished, the report is issued to the management level and a decision is reached. The decision can be unilateral or one that is negotiated between the USEPA and the regulated party. Public participation, if required, usually occurs after the assessment is completed and before the decision is made. In some programs, the law (state or federal) dictates practices, conditions, and even assumptions used in the ERA. Risks may have to be evaluated on the basis of prescribed exposure pathways, ranges of effects, populations, and parameters of the resulting numerical estimates. Some programs have criteria or limits on how the final numerical risk estimate is treated, i.e., action is taken if a hazard quotient exceeds a given value.

The Superfund program has guidelines in which a problem is described in site-specific terms, the nature of the contamination and exposure is assessed and the nature of the adverse effects or impacts is described (USEPA 1997). The risk characterization in Superfund frequently takes the form of generating hazard quotients for chemicals and individual species. More recently, risk assessments in this program have generated probability distributions as part of an effort to conduct probabilistic risk assessments. In addition, the Superfund program is one of the few USEPA offices attempting to articulate a policy for ERM (Luftig 1998). The importance of this development in Superfund has been the resulting debate and dialog that arose within the federal family and among the regulated community about whether such a policy was needed or could be developed nationally. The Superfund case study (Chapter 11) provides an example of an iterative, open ERM process where many of the elements cited in Luftig (1998) were present, despite the fact that none of those involved had an ERM guidance document to follow. The lack of and a need for a national ERM guidance document and framework were also noted by Stahl and coworkers (Stahl et al. 1999) in a recent call for the development of ERM guidance.

In other programs, the term ERA is used less and the programs have less experience to be of useful applicability here. The U.S. Department of Agriculture (USDA), for example, established the Office of Risk Assessment and Cost Benefit Analysis as part of a process for improving the analytical quality of programmatic and regulatory decision-making. As a result, the USDA is now increasing the number of risk assessments conducted, the number of risk assessors on staff, and the use of risk data in decision-making.

The generalization from these programs is that the risk-assessment process provides a substantial body of scientific and technical information to base the decision upon. This scientific information must form the core or basis for the resulting decision, but the scientific data and analyses, while necessary, are not sufficient for decision–making. The final output requires information and analysis above and beyond what the ERA provides, as has been noted previously in this book. More on this latter point can be found in Lackey (1995, 1994) and in Chapter 1.

The workshop participants recognized that the actual process of selecting options, that is, the process of making the decision, is not fundamentally unique in the realm of environmental decisions. Whether the decision is made strictly on the basis of technical analysis, public opinion polls, analytic-deliberative process, consensus, or something else, an ecological management decision follows patterns found in other decisions. What is distinct for ERM is the need to incorporate different sorts of information under such a wide range of conditions as found in ecological manage-ment situations.

The Role of Science

Use of scientific information

Most experts in the field of risk assessment agree that the scientific analysis is the foundation for the final decision. Interestingly enough, this observation was more tempered in the present context, due in part to the rich diversity of expertise and in part to the experience of all the participants. Participants had direct experience in which the scientific analysis was "necessary but not sufficient" for final decision-making. Many participants agreed that nonscientific issues have determined the nature of decisions that they were personally familiar with. These cases included allowing an activity (i.e., hunting in Tennessee) despite demonstrated damage to a resource because of long-standing cultural practices and remediating a contami-nated site at great cost and disturbance with no evidence, prediction, or concern for harm if the site remained undisturbed. Clearly the legal, social, and economic factors played critical roles in these ERM decisions.

A number of publications and analyses have examined the role of scientific and technical information in making environmental decisions. The NRC report on risk assessment in the federal government (NRC 1983), though not specifically address-ing ecological assessments, noted the underlying importance of scientific and technical information in environmental decisions. Since then, the NRC has under-taken a series of projects on risk assessment and management, including the 1994 report resulting from the Clean Air Act (NRC 1994), the National Forum on Science and Technology Goals (NRC 1996b), and the report on risk characterization (NRC

1996a). These NRC publications consistently call for rigorous scientific data gathering and analysis in assessing risks and conclude that environmental management and policy decisions must rely on a strong foundation of scientific information.

The NRC also called on the USEPA to develop a framework for ERA (NRC 1983), a process that culminated guidance for conducting ERAs (USEPA 1998). This guidance relies heavily on the scientific and technical component of ERA. In fact, the USEPA's 1998 ERA guidance is, in many ways, more developed than earlier guidance on human–health risk assessment. It calls for specifically addressing such issues as uncertainty, underlying assumptions, and alternative models in estimates of risk. In this guidance, and in subsequent efforts, the USEPA makes a concerted effort to set standards for the underlying scientific information and analysis. Experts in the field of risk and ecological risk may understand how ecological risk practice might lead the way at present. For many years, risk analysis practices focused on human health, and it has really only been in the last few years that ERA has received increased attention in federal agencies. At the same time, developments in the area of human-health risk analysis have been less dramatic and groundbreaking.

Types of scientific information

Certain information should be available and considered almost basic or standard in risk-based decisions. The following ecological information was considered standard:

- habitat and ecosystem descriptions,
- species composition,
- temporal trends in existing resources, and
- animal and plant distributions.

This information should be available for many sites or cases in question given the extent of ecological investigation conducted to date and the effort of federal programs, i.e., USEPA's Environmental Monitoring and Assessment Program, U.S. Fish and Wildlife Service wetland mapping and rare and endangered species programs, and the National Oceanic and Atmospheric Association's National Status and Trends Program (mussel watch). An increasing number of regional and large-scale management efforts have made considerable progress in this regard, including coastal Louisiana protection, Chesapeake Bay Program, South Florida Water Management, and the National Estuary Program. If site or case specific information is not available, similar or closely related data should be available and should be consulted whenever possible.

Of course, other types of scientific information should be consulted whenever it is available. Considering that the modern environmental effort is several decades old now, information exists in numerous areas—toxicology (both human and ecological), meteorology, geology, geography, and chemistry (including fate and transport of chemicals). Furthermore, new tools for data acquisition and analysis have been developed in the past few years enabling ERM to incorporate substantial amounts of information. The development of electronic databases, both on-line and portable, and geographic information systems (GIS) are two recent advances that put more information in the ERM process and in forms that should be more usable.

Uncertainty analysis

Workshop participants agreed that uncertainty analysis, whether quantitative or qualitative, is a necessary component of both risk assessment and decision-making. Morgan and Henrion (1990) identify eight different types of uncertainty that should be carefully examined as a prelude to informed decision-making involving risk, to which we add a ninth (Table 5-1). The fundamental problem, of course, is that of these nine distinctive types of uncertainty, only random error is routinely measured and reported. Worse, even random error is not reported every time that risk is estimated statistically.

Table 5-1 Types of uncertainty (modified from Warren-Hicks and Moore 1998)

Type of uncertainty	Description
Random error	-All measurement instruments are imprecise
Systematic error	-Error from improper calibration, experimental procedures, survey methods, and other factors
Subjective judgment	-Often referred to as "science policy," it is the insertion of non-scientific values held by scientists
Linguistic imprecision	-Examples include the use of non-numeric words to represent numeric quantities, and categories or sets which encompass multiple differing items
Variability	-Observed differences in realizations of random draws from a known distribution that varies in time, space, members of any population, or other dimensions
Inherent randomness	-Things which simply cannot be known or understood in practice, or perhaps even in principle
	-The classic example is the Heisenberg Uncertainty Principle, which states that the position and velocity of an atomic particle cannot be accurately measured simultaneously, because attempts to measure position affect velocity, and vice versa
Scientific disagreement	-Scientists often disagree concerning fundamental questions
Approximation	-All models are inherently incomplete specifications
Everything else	-In any taxonomy of uncertainty, there must always be a category for things we don't know that we don't know

In a recent analysis of the application of uncertainty analysis in ERA, Warren-Hicks and Moore (1998) note that these techniques have been applied in varied ways, including:

- hypothesis testing,
- estimating parameters,
- testing the distribution of a set of data,
- calculating probabilities,
- evaluating confidence limits or variability about a mean, and
- determining some measure of central tendency.

Warren-Hicks and Moore point out that the uncertainty analysis of a complex ERA may be both controversial and unusually confusing. They urge ecological risk assessors to give serious attention to the careful design of uncertainty analysis at the start of an assessment to avoid later controversy and to develop a more coherent means of displaying and explaining results.

Social and Policy Factors

Some decisions have a designated decision-maker and law that must be followed. At some point in U.S. federal regulatory and other decision-making settings, the decision stops at the desk of a person or a group or program office. This decision incorporates the information that is not scientific in ecological terms. The decision maker or team will have to use many different types of information to come up with the final outcome (the decision). The most common types of information, other than the scientific and technical information, are policy and legal requirements, economic consequences, and social (including cultural) factors (see also Chapter 3). Policies exist to incorporate all of these reflecting the values of the group, organization, or government agency; the law is a more specific manifestation of the policy. Thus, decision-makers count on the overarching statement of policy to offer guidance in cases that are new or bring unfamiliar elements together. Policy-based decisions can also be a form of avoiding the difficult choice based on an unclear scientific information base or on one that is radically different from expectations.

The social considerations of an ERM decisional process include the changes to local society and to the larger society of carrying out an activity, or not, of various options and alternatives. The NRC considered that the analytic-deliberative process was one way of addressing social factors as well as scientific analysis (NRC 1996a). In this workshop, participants raised numerous examples of how social aspects of ecological situations factored in to ERM decisions or how these social aspects should have played a role in the formal process, rather than altering the outcome after the fact. One of the more significant points made in the workshop was that the social, legal, political, economic, and cultural ramifications and aspects of ERM should be part of the initial processes that set out the analytical and management paths.

One of the important social factors is the process used to make the decision, as noted by NRC (NRC 1996a). If conducted with a flawed process, even wise decisions may fail acceptability. One known and key factor is involving the stakeholders and interested parties in the decisions (Yosie and Herbst 1998). A recent effort used a user-group and stakeholder-based method to assess the sources of stress and consequences of stress on the Fox River system from the point of view of the watershed. The interaction, facilitation, and technical research were conducted by faculty and staff at the University of Wisconsin. The process and results of the group interaction are reported in van den Belt et al. (1998) and the technical analysis in Campion (1997).

Several issues raised by this effort are relevant to ERM. Complex data and technical information needs to be presented to the wide array of stakeholders in such a way as to be understandable and usable, as it was for the Fox River. These specific participants learned new information in the process that helped in evaluations and decision-making. And the inclusion of stakeholders at an early stage was central to success. In contrast, the NRC criticized the U.S. Department of Energy for remaining apart from the stakeholders and excluding communities from remediation efforts around the nation (NRC 1997).

Communication

A growing literature on risk communication has focused more on human health issues than on ecological issues (NRC 1989). The NRC study of risk communication marks some of the important general aspects of this issue. Some of these are the need to communicate in language that is understandable to the audience and the need to communicate in a voice or tone that is believable. Other aspects of risk communication are addressed by NRC and have since been complemented by journal articles and reviews.

At the workshop, the participants recognized the need for open communications at more than one level. The first level was among the interested and affected parties who have been and will remain involved in the life of the project. These are also the parties who will in all likelihood be part of the decision in a formal fashion through public participation or by dealing with an official decision-maker after the decision is implemented. This step is a necessary part of the process and successful communication at this stage can mean the success or failure of the project or the implementation. Interested and affected parties may be identified in a number of ways, including:

- past lists of participants in related activities,
- business groups,
- citizen groups,
- public comment requests or public hearings,
- public notices, and
- personal contacts.

The importance of conducting public communication properly at the outset of the process has been addressed in other publications (NRC 1996a, 1989). Other reports, as well as the participants, emphasized the point that future trust and commitment depend on initial notification. If parties or individuals do not feel welcome, believed, heard, or valued, then their participation and trust in the process and the final decision is threatened.

The other level of communicating is with the larger constituencies, including the political entities that make larger related decisions. Communication with the greater community, including the political parties and entities, is a necessary step regardless of who is conducting the risk assessment and the decision-making. Decisions on ecological risk issues have to be viewed as having the interest, attention, and involvement of some portion of the public as well as governments.

Many ERM decisions affect public resources or natural resources in the public domain. According to law, natural resources are under the trust of the state government and other natural resources (migratory birds and rare and endangered species) are protected by a federal agency. Thus, regardless of the initiator, actions affecting these resources will touch a government agency.

One of the central processes in risk management is communicating with the affected parties and the public regarding problems and issues facing a decision-maker (see also Chapter 3). Workshop participants agreed that risk communication is such a central and pervasive part of the risk–management process, starting from the first step of identifying a problem through the final step of implementing a decision, that it would be inappropriate to establish communication as a separate step; communication must occur throughout. This point is also underscored in the six-step risk management framework developed by the Presidential/Congressional Commission on Risk Assessment and Risk Management (1997). In that framework, stakeholder engagement — that is, the iterative communication about perceived risks, concerns, and values — lies at the center of the entire process, from establishing the problem and defining its context through the evaluation of whether risk–management actions have produced the desired results.

This necessity for continual communication is seen clearly when one considers the two pivotal ERM questions discussed by conference participants: what should we protect, and how much should we protect it? On the face of it, these questions are not the kind to be tackled solely by experts. Rather, they are questions about what citizens care about, how much they care about it, and what tradeoffs they are willing or unwilling to make to achieve their ecosystem protection goals. Thus, even at the portal of an ERA, before entering into the complex steps of analysis and characterization, managers must engage in deliberations with the public about what fundamental values will guide the decision-making process. While they may sometimes be preordained by statute, at other times they will not. Successful completion of the decision-making process will hinge on successful communication with stakeholders.

Considering how central risk communication is to decision-making, it behooves the risk manager to understand clearly what risk communication is. In its 1989 report, *Improving Risk Communication*, the NRC defines risk communication in the following way:

> Risk communication is an interactive process of exchange of information and opinion among individuals, groups, and institutions. It involves multiple messages about the nature of risk and other messages, not strictly about risk, that express concerns, opinions, or reactions to risk messages or to legal and institutional arrangements for risk management.

NRC explicitly rejects the traditional notion of risk communication as the delivery of one-way messages from experts to nonexperts. Successful risk communication, according to NRC, "raises the level of understanding of relevant issues or actions and satisfies those involved that they are adequately informed within the limits of available knowledge."

As suggested in the ERM framework developed by conference participants, one of the basic questions about an ERM decision is whether it is routine or non-routine. Routine decisions, i.e., pesticide registrations, seldom demand risk communication beyond that which is necessary to meet established data submission requirements and to clarify technicalities. Non-routine decisions, which can range from the slightly contentious to the extremely controversial, demand an entirely different level of communication.

When appropriate, risk communication's central role means that risk managers must communicate with a variety of different types of people; these include scientists, analysts, activists, industry representatives, community leaders, journalists, and citizens who simply want to express their concerns and understand clearly what decision makers can tell them about risks they may face.

Because decisions rest on information of various sorts (legal, political, cultural, scientific, and economic), it would be useful for risk managers to identify their information needs through a process called "gap analysis" (Chapter 4). In this process, a manager asks the following questions: What do we know? How certain is what we know? Who knows it? What do we still need to know? Who cares and what do they want? These questions are asked about legal, political, cultural, social, scientific, and economic information. Through this process, a manager can determine if enough information exists to develop management options; if not, more information can be gathered; if so, options can be developed in conjunction with various experts and interested parties. Options development requires communication regarding decision criteria. From the list of options, managers and stakeholders can screen those warranting further analysis.

As with any good decision, an ERM decision must be clear about what information is driving the decision. It must be transparent about what is known, what is suspected but uncertain, and what is unknown. These and other critical points are made in the USEPA's seven useful rules for risk communication, which might serve to guide risk managers in developing a communication plan early in the decision-making process (Chapters 1 and 3).

Decision Implementation

The workshop members agreed that decisions must be implemented if programs and plans were to have any meaning, although all knew of situations in which such was not the case. Step 6 in the ERM framework addresses implementation (Chapter 2). The discussion was based on the assumption that implementation is carried out as the result of a management plan to follow-through on the risk-based decision.

Effective implementation requires a management system and structure with an appropriate level of reporting, accountability, and flexibility. These issues are addressed in management literature and were not the focus of the discussion in this workshop. Rather, the discussion focused on how to ensure implementation of ERM decisions in a way that seems to combine incompatible, yet necessary characteristics. The implementation needs to be sufficiently flexible to accommodate case specific conditions and to not be overly rigorous.

One of the most important aspects of the implementation phase was seen as the need to stay focused on the results of the project or plan. Risk management actions, like most other management actions, are undertaken to make a positive difference and require more than data gathering. Just as the initial phase of the assessment needs to focus on making the decision science-based, the implementation plan and actions need to focus on making a difference through actions.

Effective implementation needs the cooperation and often consent of more than one party to the action; the public, larger constituencies, and sister agencies may all have a role to play in ecological management. The workshop participants felt that this issue was perhaps more applicable to ecological than human-health issues because of the size and the extent of many ecological cases and issues. Most participants had experience with ecological decisions that involved multiple federal and/or state agencies (and some international agencies), public and private individuals and organizations, conflicting objectives for ecological and human health protection (sometimes both), and other complicating issues. Migratory species cover multiple jurisdictions and recovery plans must have the cooperation of all the entities responsible for a jurisdiction. The larger and more ambitious the project, the more important cooperation is among entities in the implementation. Two excellent examples are the Chesapeake Bay Program (see deFur 1997) and the recent program to prevent coastal wetland loss in Louisiana (Coalition to Restore Coastal Louisiana 1999). In both cases, several federal agencies were required to cooperate and work closely with local and state governments, citizens, and private business. The successes in these cases may be due in part to the ability to work among interested parties, just as the difficulties in working together is likely a limiting factor in the successes thus far. In these cases success is defined as having an open, collaborative process that leads to decisions that involve the public as well as public and private groups. One clear common feature of these programs is the leadership of federal action; both of these efforts have Congressional funding. This gives both a mandate and an incentive.

The issue of implementation may seem an obvious one, and for many scientists it is. Scientists seldom design and fund research and then do not carry out the research and publish the results. But recent reports have noted that implementation can often be the weak link in the chain of ERM (NRC 1999; Presidential/Congressional Commission on Risk Assessment and Risk Management 1997). One of the ways that implementation can be made more effective is to include program evaluation in the overall ERM design. The program evaluation piece is intended to gather information on effectiveness and efficiency and to track duties and responsibilities. As with other topics raised here, program evaluation is a field in itself, with its own rich literature base. Readers are also referred to a recent SETAC workshop and publication (Foran and Ferenc 1999) that addresses the important issues in making and evaluating risk management decisions for multiply stressed systems.

Tracking and Evaluation

Participants at the workshop recognized the need to measure success, or failure, and gather the information necessary to make changes in programs and activities. The framework includes the step of measuring as a critical aspect of the ERM process. Here is where many experts felt the present system failed. The monitoring, or tracking and evaluation, may be the last item on the budget list, the part of the project that falls to another office, division or even agency, or may be the responsibility of entirely new staff hired after the ERM process.

Monitoring the effectiveness of environmental programs, including the decisions for ecological risk situations is a critical element of the entire process. The participants agreed completely with the observation that progress is only measured if the conditions are monitored. Many years of experience in public and private programs (restoration as well as mitigation) demonstrate that individuals and organizations keep track of those activities that are important and that what is not monitored gradually becomes unimportant. The federal requirement to monitor and report emissions and releases under federal right-to-know legislation and the tracking of wetland loss in Louisiana (Coalition to Restore Coastal Louisiana 1999) are two examples of programs that showed progress once the efforts were made to keep track of program results.

A recent NRC report on watersheds (NRC 1999) noted the need to focus considerable efforts on activities such as watershed management that have not been carried out as today. Many changes in environmental science and management now address problems (radon) or use approaches (GIS, watershed management) not even considered until a few years ago. For this reason, as well as many others, it is critical to gather information on how well ERM decisions result in environmental improvements such as restoration of habitats, recovery of species, or reduction of toxic chemical levels.

Monitoring is also the only way to implement any iterative actions or course corrections in environmental programs, thus monitoring is one of the important aspects of effective programs. The group was in agreement that ERM decisions will always be based on less than complete knowledge and on ecosystems (or their components) that are changing in time and space. Thus, monitoring is necessary to track the nature of the species or ecosystem elements that are the object of the management activities. Management plans must put in place an effective monitoring program that will provide the feed back necessary to determine what, if any changes or corrections are needed in the implementation plan (see Foran and Ferenc 1999).

Summary

In this chapter a number of important issues involved with ERM decision that have been discussed, along with the important elements that are needed in ERM framework. Because decisions are impacted by a host of influences it is important to have consensus-driven elements of a framework or decision-making process in hand that drives the decision maker to a timely, effective decision. The absence of such a framework results in decisions that can be drawn out, highly controversial, poorly understood, inconsistent, or even incorrect. Influences on the decision maker and the decision itself include scientific data and debate, public policy, and economics. Once the decision is made it is also important to implement it and track its effectiveness. Thus the framework needs to include a mechanism for this implementation and tracking, and where necessary, to offer sufficient flexibility so that changes to the implemented action can be effected if needed. It is also important to have clear, open, and direct communication with stakeholders and others who are directly or indirectly impacted or interested in the decision. These individuals or groups need to provide their input and concerns to the decision maker in an efficient and effective form.

References

Barnthouse LW, Biddinger GR, Cooper WE, Fava JA, Gillett JH, Holland MM, Yosie TF, editors. 1998. Sustainable environmental management. Society of Environmental Toxicology and Chemistry (SETAC) Pellston Workshop on Sustainability-Based Environmental Management; 25-31 August 1993; Pellston, MI. Pensacola FL: SETAC. 102 p.

Campion J, editor. 1997. The upper Fox River basin: Analysis of demographic composition, public goods and natural resources. Green Bay WI: University of Wisconsin. 123 p.

Coalition to Restore Coastal Louisiana. 1999. No time to lose. Baton Rouge LA: Report by the Coalition to Restore Coastal Louisiana. 54 p.

deFur PL. 1997. Ecological risk assessment: Lessons from the Chesapeake Bay. *Amer Zool* 37:641-649.

Foran JA, Ferenc SA. 1999. Multiple stressors in ecological risk and impact assessment. Society of Environmental Toxicology and Chemistry (SETAC) Pellston Workshop on Multiple Stressors in Ecological Risk and Impact Assessment. 13–18 September 1997; Pellston, MI. Pensacola FL: SETAC. 100 p.

Lackey RT. 1994. Ecological risk assessment. *Fisheries* 19(9):14–18.

Lackey RT. 1995. The future of ecological risk assessment. *Hum Ecol Risk Assess* 1(4):339–343.

Luftig S. 1998. Draft ecological risk management principles for Superfund sites. Washington DC: OSWER Directive 9285. 7–28 p.

Morgan MG, Henrion M. 1990. Uncertainty: A guide to dealing with uncertainty in quantitative risk and policy analysis. Cambridge: Cambridge University Press.

[NRC] National Research Council. 1983. Risk assessment in the federal government: Managing the process. Washington DC: National Academy Press. 191 p.

[NRC] National Research Council. 1989. Improving risk communication. Washington DC: National Academy Press. 332 p.

[NRC] National Research Council. 1994. Science and judgement in risk assessment. Washington DC: National Academy Press. 651 p.

[NRC] National Research Council. 1996a. Understanding risk. Washington DC: National Academy Press. 249 p.

[NRC] National Research Council. 1996b. Linking science and technology to society's environmental goals. Washington DC: National Academy Press. 530 p.

[NRC] National Research Council. 1997. Building a foundation for sound environmental decisions. Washington DC: National Academy Press. 87 p.

[NRC] National Research Council. 1999. New strategies for America's watersheds. Washington DC: National Academy Press. 311 p.

Pittinger CA, Bachman R, Barton AL, Clark JR, deFur PL, Ells SJ, Slimak MW, Stahl RG, Wentsel RS. 1998. A multi-stakeholder framework for ecological risk management: Summary from a SETAC workshop. Summary of the SETAC Workshop on Framework for Ecological Risk Management; 23–25 June 1997; Williamsburg, VA. Pensacola FL: SETAC. 24 p.

Presidential/Congressional Commission on Risk Assessment and Risk Management. 1997. Framework for environmental health risk management. Washington DC: Office of the President. 64 p.

Reinert KH, Bartell SM, Biddinger GR, Webb E, Borison A, Cacela D, Christensen S, Evans M, Hunter R, Jones H, Kaplan P, editors. 1998. Ecological risk assessment decision-support system: A conceptual design. Society of Environmental Toxicology and Chemistry (SETAC) Pellston Workshop on Ecological Risk Assessment Modeling; 23–28 August 1994; Pellston, MI. Pensacola FL: SETAC. 98 p.

Stahl RG, Barton A, Bachmann R, Clark JR, Ells S, Pittinger CA, Slimak M, Wentsel R. 1997. Concepts in ecological risk management: A discussion document for the SETAC Ecological Risk Management Workshop; 24-27 June 1997; Williamsburg, VA. Pensacola FL: SETAC. 46 p. (unpublished).

Stahl RG, Pittinger CA, Biddinger GR. 1999. Developing guidance for ecological risk management. Risk Policy Report. 14 May 1999. p 38–40.

Slovic P. 1993. Perceived risk, trust and democracy. *Risk Anal* 13:675-681.

[USEPA] U.S. Environmental Protection Agency. 1994. Managing ecological risks at EPA. Issues and recommendations for progress. Washington DC: USEPA. EPA-600-R-94-183.

[USEPA] U.S. Environmental Protection Agency. 1995. Ecological risk: A Primer for risk managers. Washington DC: USEPA. EPA-734-R-95-001. 36 p.

[USEPA] U.S. Environmental Protection Agency. 1998. Guidelines for ecological risk assessment. Washington DC: USEPA. EPA-630-R-95-001F.

[USEPA] U.S. Environmental Protection Agency. 1997. Ecological risk assessment guidance for superfund: process for designing and conducting ecological risk assessment. Interim final. Washington DC: USEPA. EPA-540-R-97-006.

Van den Belt M, Harris HJ, Wenger R. 1998. Mediated modeling project. An integrated scoping model of the upper Fox River basin. Green Bay, WI: University of Wisconsin- Green Bay. 43 p.

Warren-Hicks WJ, Moore DRJ, editors. 1998. Uncertainty analysis in ecological risk assessment. Society of Environmental Toxicology and Chemistry (SETAC) Pellston Workshop on Uncertainty Analysis in Ecological Risk Assessment; 23–28 August 1995; Pellston, MI. Pensacola FL: SETAC. 277 p.

Yosie FY, Herbst TD. 1998. Using stakeholder processes in environmental decisionmaking. Washington DC: Ruder Finn.

CHAPTER 6

Using Economic Principles for Ecological Risk Management

Richard B. Belzer

Introducing the Economics-Based Framework for Making Ecological Risk Management Decisions

Chapter 5 set forth alternative frameworks for ecological risk management (ERM), highlighting several recent examples produced by distinguished panels. Members of the workshop generally were familiar with and supportive of these various frameworks, though perhaps frustrated with their limited record of success in application. Thus, while there was little argument that "risk-based decisions should take into account the full range and diversity of quantitative and qualitative information on social, legal, economic, and scientific aspects of the decision" (Chapter 5), few ideas were presented that offered systematic ways of integrating this information in a way that led toward consistent and satisfying outcomes.

Chapter 5 also sets forth three questions that a desirable ERM framework should answer: What ecological resources should be protected? How much protection should each resource receive? How should the desired level of protection be achieved?

While each of the alternative frameworks presented in Chapter 5 might be capable of answering these questions in any particular case, it is not clear that any of them can achieve consistent, scientifically credible outcomes across cases. This problem arises from the absence within these frameworks of clearly articulated procedures and methods for weighing and evaluating scientific and nonscientific information.

In contrast, an economic framework for ERM also was presented at the workshop, which was derived in large part from the intuition of ecological scientists and non-economists present. This framework, which was based on benefit-cost analysis, sets forth a systematic method for evaluating and comparing the consequences of alternative decisions. It is explicitly practiced in government and business, and it can be routinely observed in individual decision-making. Despite its ubiquity, however, benefit-cost analysis is frequently misunderstood and distrusted by many noneconomists.

Risk Management: Ecological Risk-Based Decision-Making. Ralph G. Stahl, Jr. et al., editors.
©2001 Society of Environmental Toxicology and Chemistry (SETAC). ISBN 1-880611-26-0

The objective of benefit-cost analysis is to determine which option among an array of policy or programmatic choices achieves an efficient allocation of resources. In this context the term "resources" refers not just to ecological resources, but also to the myriad assets of a complex physical and social system. Efficiency is obtained when resources are utilized in a manner that no trade, exchange, substitution, or departure can increase net social welfare. Thus, an efficient allocation is one where the maximum possible excess of benefits over costs is obtained. Cost-effectiveness analysis is a limited version of benefit-cost analysis in which either the maximum value is obtained from a fixed quantity of resource or a fixed quantity of value is obtained from the minimum expenditure of resources. This version is frequently practiced where benefits can be quantified but it is difficult or impossible to value them or where the desired goal is ambiguous.

A common source of confusion is the difference between economic and financial benefits and costs. Economic benefits are improvements in welfare, whether they accrue to individuals, families, or communities, and are measured in terms of added consumption of goods or services. The term consumption is frequently misunderstood to imply the use of material resources. The protection of an ecological resource is a form of consumption in economic terms as long as the act of protection improves societal welfare. Economic costs consist of reductions in real resources. These resources may be physical and tangible (i.e., routine consumption of goods and services, clean air and water, or an ecological system) or intangible (i.e., abstract goods like freedom and justice). The defining factor is whether the item is scarce, meaning that it is not available in unlimited supply.

In contrast, financial benefits and costs are purely pecuniary. They occur because the transactions that yield social benefits and costs are largely performed through currency and other media of exchange. In regulatory settings, a financial cost arises when a regulated entity is compelled to spend resources to comply with some regulatory standard, monitor operations, submit paperwork to the government, or disclose information. These expenditures typically correspond to economic costs because real resources are consumed along the way. However, certain expenditures are financial costs but not economic ones, and for many economic costs, there are no financial expenditures.

A clear example of a cost that is financial but not economic occurs when a regulated entity must acquire equipment or services from a particular supplier who enjoys market power, perhaps because of the regulatory requirement. Expenditures in excess of marginal cost may constitute monopoly profits to the supplier. While these are real financial costs to regulated entities, economists count them as pecuniary expenditures because they represent resources that are transferred from buyers to sellers but are not consumed. An example of a real economic cost that has no financial expenditure associated with it occurs whenever a regulatory agency prohibits a product, service, or activity. In each case, real resources are consumed because the benefits afforded by the opportunity to purchase the banned product or service or engage in the forbidden activity is lost.

Oftentimes, keeping economic and financial effects straight turns out to be a difficult task. For ecological scientists and other noneconomists, the critical lesson is to avoid trusting one's own intuition as a substitute for professional economics insight. Just as economists should leave ecological science to ecologists and assist only with helping to structure ERM alternatives and value alternative outcomes, noneconomists are well advised to steer clear of the valuation component of ERM and focus instead on ecological science.

At least two federal environmental laws have long required the balancing of benefits and costs in regulatory decision-making. The registration and regulation of pesticides under the Federal Insecticide, Fungicide, Rodenticide Act (FIFRA) requires the U.S. Environmental Protection Agency (USEPA) to use benefit-cost analysis in making pesticide registration decisions. Under the Toxic Substances Control Act, the USEPA may regulate or even ban a chemical if it poses an unreasonable risk. Historically, the threshold for showing an unreasonable risk has been a demonstration that a specified regulatory intervention offers greater social benefits than social costs.[1]

Apart from specific environmental laws, there is an overarching requirement that executive branch agencies perform benefit-cost analyses for all proposed or final regulations that may impose costs of $100 million or more in any one year. This requirement has been recorded in successive presidential Executive Orders since 1981. The Office of Management and Budget reviews all significant regulatory actions and examines the validity and reliability of agencies' benefit-cost analyses as part of the deliberative process prior to regulatory decision-making.[2]

In describing how the standard economic framework can be applicable to ERM, it is essential to emphasize at the inset what the economic framework is not about. In particular, it is not about how to elevate the consideration of costs above concern for the natural environment or devising a means of avoiding the commitment of scarce resources for environmental protection. Probably without exception, environmental economists dedicated to ecological values would argue that these myths or misunderstandings represent the single most important impediment to serious protection of natural environments, because they have so thoroughly discolored honest intellectual debate that many noneconomic scientists reflexively distrust the economic paradigm. What economists know instinctively (and ecologists struggle to admit) is that preservation entails tradeoffs.

---

[1] See Cropper and Oates 1992. The Food Quality Protection Act (FQPA) of 1996 amended the Federal Food, Drug and Cosmetic Act (FFDCA), which is the statutory authority under which USEPA has set tolerances for pesticides used on produce. Prior to FQPA, USEPA was required to implement conflicting statutory directives — FIFRA, which contains the risk-benefit balancing language described in the text, and FFDCA, which because of the infamous Delaney Clause required USEPA to ban any pesticide shown to cause cancer in man or animal at any dose. FQPA repealed part of the Delaney Clause but significantly changed policy for setting tolerances. The net effect of these changes is still unclear, in large part because regulations under FQPA have not yet been litigated. Nevertheless, it is conceivable that pesticides that would not have been substantially regulated at all under FIFRA because benefits substantially exceeded risks will be effectively banned under FQPA.

[2] See Executive Order 12866 (58 Federal Register 51735, October 4, 1993); Executive Order 12291 (46 FR 13193, February 17, 1981).

A Short History of the Economics Literature Related to Ecological Risk Management

For many noneconomists it can be something of a revelation to discover that economists have studied environmental problems so extensively. Indeed, a huge part of the relevant economics literature predates the modern scientific study of ecology.

The economics literature on conservation began in earnest during the Progressive Era and its rhetoric strikingly resembles what can be routinely encountered today among strident environmentalist tracts. Ruminations on the exploitation of natural resources by Professor L.C. Gray of the University of Wisconsin are representative of the period, showing beyond any doubt that economists of the time held no brief in support of development:

> The habit of exploitation, once fixed, is so strong that individuals will frequently continue their wasteful practices after they become uneconomic even from the point of view of individual self-interest (Gray 1913).

Indeed, a common thread in economic writings of the period is sympathy towards governmental ownership or control of the means of production; Professor Gray himself was an avid price-fixer who celebrated state control over markets.

These sympathies began to wane as economics slowly shed its predilection for moral philosophy uninformed by scientific theory and empiricism and developed policy-neutral theories of individual, firm, and market behavior. Economists have tackled problems related to the depletion of exhaustible resources (e.g., Devarajan and Fisher 1981; Dasgupta and Heal 1979; Hotelling 1931), the optimal use of renewable resources (e.g., Baumol and Oates 1975; Ciriacy-Wantrup 1952; Kotok 1945), and the preservation of natural environments (e.g., Fisher and Peterson 1976; Krutilla and Fisher 1975; Arrow and Fisher 1974; Krutilla 1967). An entirely new and initially controversial branch of economic theory and analysis relevant to ERM sprang to life forty years ago (e.g., Coase 1960, made popular by Dales 1968) which instead focused on the assignment of privately held and publicly enforceable property rights. These controversies have faded away over the years, and it is now generally accepted within the profession that private property rights, even to environmental amenities and ecological resources, offer an extremely effective means of preservation. The collapse of the Soviet Union has revealed ample evidence that the forced abandonment of private property rights in favor of state ownership and control may be the best possible prescription for environmental disaster.[3]

The applicable lessons of the economics literature have not been codified in federal law, policy, and regulation reflects the fact that the relevant laws were written a

---

[3] Pryde 1983 concludes that environmental quality depends on industrial activity per se and not the nature of the economic system, despite providing persuasive evidence to the contrary obtained from sanitized Soviet sources. Kryuchkov 1993 describes an "ecological apocalypse" in the Siberian arctic from oil and gas exploration, mining, and steam-generated electric power, a situation far more grave than any documented under Western capitalism.

generation ago and thus do not reflect current economic (or scientific) thinking. For example, the admirable desire to preserve biological diversity expressed in the Endangered Species Act of 1973 was not accompanied by practical and transparent methods to accomplish this goal in a manner that was consistent with other social values. In an article devising a systematic way to model biological diversity, Weitzman (1992) sets the stage succinctly:

> Often there is an implicit injunction to preserve diversity because it represents a higher value than other things, which by comparison are "only money." Yet the laws of economics apply to diversity also. We cannot preserve everything. There are no free lunches for diversity. Given our limited resources, preservation of diversity in one context can only be accomplished at some real opportunity cost in terms of well-being forgone in other spheres of life, including, possibly, a loss of diversity elsewhere in the system.

Another important factor explaining delays in the incorporation of economic methods is the extent to which federal agencies charged with implementing the ecological protection and preservation policies of the last generation are wedded to the scientific and economic constructs from which these policies were derived. Further, federal regulatory agencies have generally been loathe to adopt the economic paradigm in part because it can raise painful doubts about both the efficiency and effectiveness of agency performance.

The remainder of this chapter provides an elementary treatment of certain fundamental principles of economic theory applied to ERM, theory that can be found in any intermediate level undergraduate textbook in microeconomics. Given the huge gap between the current state of the economics literature and the level of economic understanding among ecological scientists, any more intensive or sophisticated treatment here is clearly inappropriate. The use of elementary models is intended to begin the process of bridging this gap in order to advance the day when ecologists and economists can effectively collaborate on joint efforts in environmental protection.

What Ecological Resources Should be Protected?

Protecting the natural environment has value because people, both as individuals and as members of society, are willing to pay for the benefits obtained from environmental protection. Whether these benefits take the physical form of biological diversity, scenic vistas, wilderness, or urban open space matters less than the fact that people truly are willing to pay to obtain them. The concept of willingness-to-pay has a rich and complex literature in economics, but it may be simplified here to mean that people (and societies) are willing to sacrifice the benefits they could receive from other valuable things in order to obtain those things they value more highly.

In practice, ecologists may not be happy with popular preferences that are reflected in individuals' observed willingness-to-pay. For example, lay citizens may place disproportionately high values on species, locations, and ecological systems that appear visually attractive or are otherwise engaging. Because valuation is founded on personal and societal tastes, these popular preferences can be shaped somewhat to better conform to scientific knowledge and understanding. At the same time, preference-shaping through scientific education differs from psychological manipulation, which cannot be ethically justified. Nor is there a scientific or ethical basis for supplanting individuals' preferences with others, e.g., the preferences of ecologists, simply because education is too expensive, time-consuming or ineffective.

In a simple economic model of an ecological resource, the quantity of that resource is a function of price, which is the same thing as willingness-to-pay. There will be an optimal quantity of this ecological resource that depends on both its marginal value and the marginal cost of acquiring or protecting it.

Figure 6-1 illustrates the benefit side of this relationship. Total benefits of an ecological resource, denoted by *E*, rise with the quantity of that resource provided. While total value rises with quantity, it does so at a decreasing rate reflected in the convex shape of the curve. Convexity is assumed because an ecological resource, like other things people value, becomes less valuable per unit as the total quantity of that resource rises. The first thousand acres of wetlands added to the Everglades will be more valuable per acre than the 1,000th thousand acres. Thus, marginal benefit, which is simply total benefits divided by quantity at each point on the (horizontal) quantity axis, declines as the quantity of *E* increases.

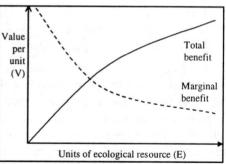

Figure 6-1 Benefits from protecting an ecological resource

Figure 6-2 displays the cost side of the ledger. The total cost of acquiring or protecting an ecological resource rises with the quantity of the resource acquired or protected. More impor-

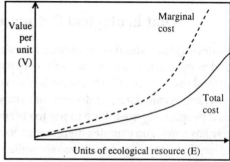

Figure 6-2 Costs of protecting an ecological resource

tantly, total cost rises at an increasing rate. This occurs for both technological and economic reasons. Technically, the acquisition, protection, or restoration of an ecological resource, becomes increasingly difficult as the available choices of ecological resources remaining diminish. Economically, an increasingly large amount of benefits from other goods and services must be sacrificed to obtain these additional units. Costs are simply the benefits of things foregone.

Thus, the marginal cost of acquiring, protecting or restoring an ecological resource rises at a rate faster than total cost. Like its benefit analogue, marginal cost is defined as total cost divided by quantity at each point along the (horizontal) quantity axis. The first units of an ecological resource (e.g., a wetland that delivers local environmental services) are the least costly to acquire or protect. As the number of units acquired or protected rises, the cost of securing each additional unit rises. The 1,000th acre of new Everglades' wetland acquired will cost more than the 999th unit, which itself will cost much more than the first unit. Expanding the size of the Everglades such that it encroaches upon the city limits of Miami would entail massive foregone benefits. (This does not imply that turning Miami into a wetland would have no benefits at all. It says only that the cost of doing so, measured in terms of the benefits foregone, would become extremely great. This suggests that Everglades' expansion elsewhere at the margin would likely have greater net benefits and it hints about the conclusion; there is an optimal size for the Everglades).

Cost is not necessarily measured in monetary units, such as dollars, rather, cost is correctly understood as the value of benefits from other goods and services that must be foregone to have the ecological resources in question. Thus, if the ecological resource is a wetland, then cost is measured as the total value people place on the goods and services that must be sacrificed to devote the freed resources to wetland protection.

Ecologists (and other noneconomists) often struggle with this graphical presentation for at least two reasons. First, economics appears to be alone in having reversed the usual convention and placed the dependent variable in the relationship along the horizontal rather than the vertical axis. There are sound reasons for this convention in economics that need not be described in detail here. Simply recognizing that the graphical representation in economics is "backwards" may be sufficient to help overcome this confusion. It also helps remind noneconomists that the usual convention of placing the dependent variable on the vertical axis is arbitrary, not inherently superior.

The second reason why ecologists struggle with this formulation is substantive rather than symbolic. Ecologists are quick to point out that all units of an ecological resource are not equal. Certain acres of wetland will tend to be critical for the production of environmental services, whereas other acres may be useful, but less important. This observation is correct and actually illustrates the fundamental reason why economic reasoning can aid in structuring the ERM problem. Because all physical units of an ecological resource are indeed different, it becomes the task of ecological scientists to rank them, preferably on a metric rather than an ordinal scale, for the purpose of developing an instrumental definition of the units of the resource. Frequently this may take the form of identifying and quantifying the environmental services that ecological resources produce and assigning to each physical unit the specific quantity of environmental services provided. True complexity arises in taking account of myriad interactions across adjacent or related physical units, nonlinearities in the level of environmental services provided at different levels of protection, and the complicated (but unavoidable) problem of assigning weights to each environmental service (weights that reflect the consensus view among ecological scientists concerning the relative ecological importance of different services).

The task of devising a scientifically defensible measure of ecological services is surely a daunting one, but the need to devise it is not a unique requirement of the economic paradigm. Regardless of the decision-making framework chosen for ERM, ecological risk assessors cannot expect to make scientifically intelligible and consistent recommendations to ecological risk managers unless they do so. Where the economic paradigm is unique is that it provides a framework for actually using this information in decision-making.

How Much Protection Should Each Resource Receive?

Figure 6-3 illustrates what economists call the optimal or efficient quantity of an ecological resource as well as optimal or efficient price or value per unit. It displays the relationship between the marginal value of an ecological resource and the marginal cost of providing it. The optimum occurs where marginal benefit and marginal cost intersect and thus are equal. This optimum level of the ecological resource is labeled E^*, and the unit price of V^* reflects the amount people are willing to pay for the last unit of the resource necessary to achieve E^*. For any level of the ecological resource less than E^*, the marginal benefit of expanding it (or protecting more) exceeds what must be sacrificed in order to obtain it, and thus, more of the resource should be produced (or protected). However, for any level of the ecological resource greater than E^*, the marginal benefit it yields is less than the marginal cost of providing (or protecting) it. At these levels, too much is being sacrificed to obtain these resources. Society would be better off if less of the ecological

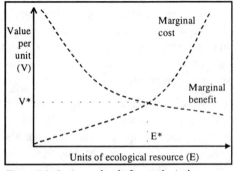

Figure 6-3 Optimum level of an ecological resource

resource were obtained or protected and other things of value were sought instead. Paraphrasing Weitzman (1992), some of these "other things" may be other ecological resources.

The fact that an optimum exists says nothing about the level at which the optimum is actually located. For critical or unique ecological resources that have few, if any, close substitutes, the optimum is likely to be located at a fairly high level. The environmental services obtained from these ecological resources may be difficult or perhaps impossible to obtain elsewhere. Conversely, for rather ordinary ecological resources that have many excellent substitutes, the optimum level may be quite low. Thus, this conceptual framework is scrupulously neutral regarding the actual outputs of decision-making. It merely provides a structure for answering the second fundamental question facing ecological risk management.

In practice, of course, it may be extremely difficult to precisely estimate either E^* or V^*, the optimum quantity and unit value of ecological resources, respectively. For many real-world policy problems, a satisfactory objective is to get close to the optimum, or, less optimistically, to avoid taking actions that send us further away from it in either direction. Figure 6-4 divides the alternative outcomes into three categorical levels. Below some arbitrary and subjective threshold such as E_L, the level of ecological resources may be judged too low. The difference between marginal benefit and marginal cost is understood to be positive (i.e., $MB|E_L - MC|E_L > 0$) though the magnitude of this difference is uncertain. Similarly, above some arbitrary and subjective upper threshold E_U, the level of ecological resources is widely viewed as excessive. The difference between marginal benefit and marginal cost here is understood to be negative (i.e., $MB|E_U - MC|E_U < 0$) though the magnitude of this difference also is uncertain.

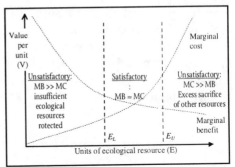

Figure 6-4 Satisfactory level of an ecological resource

Obtaining any point outside of this middle ground could be judged unsatisfactory based on qualitative analysis or simple intuition, and a satisfactory policy would achieve a level of ecological resources between these lower and upper thresholds. In this range, marginal benefits and marginal costs will be similar in magnitude (or at least not demonstrably different), assuring that a reasonable level of the ecological resource is obtained.

Critical for applying this framework is the development of methods for valuing ecological resources. This process has lagged way behind the development of methods for valuing human health risks for a number of reasons. First, ecological risk is less salient among members of the public than human health risk. People generally have a higher regard for people than they do for other life forms. While some would criticize this anthropocentric bias implicit in the economic paradigm, it is a bias present in all ERM frameworks – even those that value nonhuman biota above humans, for it is still humans who assign the relative values. The difference is that the economic paradigm relies upon aggregate expressions of individual values whereas other frameworks adopt values derived from various elites. The former mirrors democratic participation in the valuation enterprise, whereas the latter do not.

Second, people have numerous ways to express their willingness-to-pay to reduce health risks through normal marketplace transactions. While there are fewer market opportunities to express willingness-to-pay for ecological resources, such opportunities are by no means hard to find. Individuals express these values when they act to preserve or enhance natural environments under their control, or when they support organizations that acquire sensitive tracts of land. Valuation is considerably easier for human health risk, but the valuation of ecological risk is not impossible.

Third, the valuation enterprise has encountered considerably greater resistance among ecologists. Unlike human health risks, for which there is a long tradition of valuation in medical practice, there are few scientific or cultural markers for valuing ecological risks. An intensified and concerted research program is necessary to bridge this gap before the economic model of ERM decision-making will gain wide acceptance. That said, continued neglect of this research need may consign ERM to a variety of nonscientific frameworks that do not advance the cause of protecting ecological resources, except by chance.

Adding Diversity among Ecological Resources

So far, we have addressed the matter of how much protection a specific ecological resource should receive. In principle, the model described above can be readily modified to account for multiple ecological resources, and thereby yield an answer to the question of how much protection should be devoted to ecological resources in the aggregate. The optimum price and quantity will differ for each resource. However, if the quantity axis is transformed from raw units of differing ecological resources into a generalizable index of ecological services, then the model yields a vitally important result: at the optimum, between any pair of ecological resources the ratio of marginal benefits will equal the ratio of marginal costs.

A much more important source of complication is that ecological resources cannot be assumed to be always independent. Protecting one ecological resource may impose costs on a different one, or protecting some collection of resources simultaneously may yield benefits greater or less than the sum of protecting them individually. Such interactive relationships, if known or well estimated by ecologists, can be incorporated into the model with a corresponding increase in mathematical complexity. In the absence of such knowledge, however, each ecological resource problem may have to be handled as if it is in fact independent, though with perpetual care toward the prospect of interactive effects.

Tradeoffs among Ecological Resources

This model finesses the question of how an index of ecological value or environmental services might be constructed. Today, there is no generally accepted index available, and efforts to devise one typically falter because of the subjective character of any indexing system. Further research on generalized indexes surely would help, but it seems unlikely that any candidate index would prevail unless it was constructed from the ground up based on the collective expertise of ecologists who share a convergent scientific understanding. Economics has nothing to add in this regard, except for highlighting the dismal fact that advancements in ERM depend on it.

Nevertheless, the economic paradigm provides a structured means for ecologists to confront this difficult issue without necessarily having to reduce all ecological resources into ecological service building blocks. That is, rather than apply reductionist principles to the task and then reconstruct ecological resources as packages or portfolios of these environmental service building blocks, there is an alternative way to make judgments concerning the relative merit of alternative packages of ecological resources.

Figure 6-5 provides a way of conceptualizing this task. Consider the case where there are only two ecological resources, S and T, and assume for simplicity that the effort devoted to enhancing or protecting each has no effect on the other. Because S and T are mutually exclusive, each curve represents a constant total value obtained by enhancing or protecting any specified combination of the two. For example, the curve V_1 represents a constant value of ecological resources consisting of alternative combinations of resources S and T. The curve is convex to the origin because the marginal benefit of each additional unit of resources S and T, respectively, declines the more of it we already have and the more of the other resource we must sacrifice to gain it. We can have a lot of resource S and relatively little of resource T, such as at the point (s_1, t_1), or alternatively we can obtain a lot of resource T and

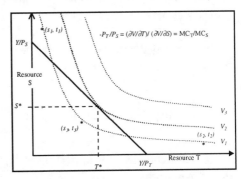

Figure 6-5 Optimum mix of two ecological resources

relatively little resource S, such as at the point (s_2, t_2). Or we could select a moderate level of both resources S and T, such as at point (s_3, t_3). We are said to be "indifferent" to these combinations because, by definition, each combination has equal (ecological) value.

The curve V_1 represents just one of an infinite number of such equal-value curves. Curves located toward the upper right represent progressively higher total values, so they are clearly preferred over curves located toward the lower left. If ecological resources could be enhanced or protected at no cost — meaning without having to sacrifice the foregone benefits of other things — then we would strive toward the highest value curve that is technically feasible to achieve.

Of course, the resources available to enhance or protect ecological resources are limited. Each incremental investment in the enhancement or protection of ecological resources means fewer resources available for other valued goods and services. The downward-sloping chord drawn between the two axes represents an arbitrarily fixed total amount of societal resources (cost) that will be devoted to enhancing or protecting ecological resources, and thus it captures the tradeoff between resources S and T. Higher (lower) chords denote larger (smaller) fixed commitments to protecting resources S and T.

For the moment, assume that this chord represents the total expenditures available for enhancing or protecting resources S and T. The linear shape of this chord implies that, as a technical matter, expenditures may be freely devoted to any mix of resources S and T. (The chord would have a stair-step shape if ecological protection were physically "lumpy.") If expenditures were devoted entirely to resource S, then the amount of resource S obtained would be Y/P_S, where Y is defined as the total expenditures available for protecting resources S and T, and P_S is defined as the unit cost (price) of protecting resource S. Conversely, the maximum amount of resource T that could be obtained would be Y/P_T, where Y retains its prior definition and P_T is the unit cost of protecting resource T.

Since higher levels of ecological protection are desirable by definition, the optimal mix of resources S and T occurs where the fixed expenditure chord is tangent to a constant-value curve. In Figure 6-5 that constant-value curve is V_2. For the fixed expenditure Y, S^* and T^* represent the particular mix of resources S and T, respectively, that achieves the greatest possible value of these ecological resources given the expenditure constraint.

Achieving the Desired Level of Ecological Risk Management

The only remaining variable in the economic framework is the total expenditures available for enhancing or protecting ecological resources. So far we have assumed that expenditures are fixed. The advantage of approaching the ERM problem this way is that it closely approximates real-world project-level decision making. Over longer time horizons and in broader contexts, however, expenditures on ecological protection only appear to be constant when in fact they are better characterized as sticky. Prior experience and practice provide powerful anchors for how we allocate resources between ecological risk reduction and other things. New information, scientific insights, and social trends, among other factors, constantly reshape individual and societal preferences such that the amount we are willing to spend waxes and wanes. The most important of these factors is likely to be wealth. The protection of ecological resources is much more highly valued among the rich than it is among the poor. Hence, general efforts to expand wealth will yield greater levels of ecological protection, whereas programs and policies that destroy wealth will reduce it.

Especially in poor communities, the level of expenditures devoted to ecological risk reduction must compete intensely with other valued objectives. During periods of economic or political turmoil, individuals will reallocate resources away from ecological risk reduction and toward food, housing, safety, and other basic human concerns. Such changes are normal and unavoidable, and they make large-scale investments in ecological risk reduction more risky — just like large-scale investments in other areas. There is always the possibility that tastes may change, making yesterday's ideal project tomorrow's white elephant. As a general matter, risk (understood here as uncertainty of outcome rather than an undesirable result) has an important effect on willingness-to-pay. That is, the natural human quality of risk aversion leads people to value a package of uncertain benefits at less than their expected value. Conversely, risk aversion also will lead them to value a package of uncertain costs at greater than their expected value. By avoiding extravagant scope and scale in the design of their projects, ecological risk managers can at least partially offset these effects.

Conclusion

The economic paradigm answers the three questions set forth at the beginning of both this chapter and Chapter 5. It does so in a manner that respects both scientific knowledge and social values, as expressed by the willingness-to-pay of individuals and the communities in which they live for the protection and enhancement of ecological resources and the reduction of ecological risks. Resources that have value should be protected, and those which have greater value should be protected more intensively. Similarly, resources that are relatively inexpensive to protect will frequently warrant greater expenditure because protecting them will add more to the aggregate value of ecological resources protected. However, both the gains from protection and the losses from foregone benefits must be balanced to maximize value. Economics provides a simple framework for understanding how this balancing act can be sustained. For any collection of resources, protect each such that the ratio of their marginal benefits equals the ratio of their marginal costs. The same reasoning applies when evaluating ecological risk reduction against other valued goods and services. For example, in a world where only ecological risk reduction and higher education matter, the optimal quantities of both can be achieved by equalizing the ratios of marginal benefits and marginal costs.

The most important lesson for noneconomists is that tradeoffs cannot be avoided. Effective ERM will not be advanced by denial. Some may raise ethical objections to the use of benefit-cost analysis (though these objections can be debated effectively), but the advancement of ethics-based objections inevitably sacrifice the very goals that prompted the objections themselves and thus raise their own ethical problems. Ecologists will accomplish much more by striving to understand and properly apply the economic framework that underlies benefit-cost analysis.

Mathematical Digression

The model described graphically in the main text can be shown mathematically using methods found in any intermediate-level undergraduate text in microeconomics. First, taking the total differential of the constant-value function, $V=V(S,T)$, yields

$$dV = \frac{\partial V}{\partial S} dS + \frac{\partial V}{\partial T} dT \qquad \text{(Equation 6-1)},$$

where the partial derivatives represent the marginal contributions to total ecological value of each resource. For any movement along a constant-value curve (such as V_2, in Figure 6-5), total value is unchanged by definition, so dV may be set equal to zero and obtain

$$\frac{\partial V}{\partial T} dT = -\frac{\partial V}{\partial S} dS \qquad \text{(Equation 6-2)}.$$

Rearranging terms

$$\frac{dS}{dT} = -\frac{\partial V / \partial T}{\partial V / \partial S} \qquad \text{(Equation 6-3)}.$$

This means that the slope of a constant-value curve, dS/dT, equals the (negative) ratio of the marginal contributions each resource makes to the total value of ecological resources protected.

The standard technique for deriving the optimal mix of two goods such as resources S and T involves crafting the following Lagrangian expression:

$$L = V(S,T) + \lambda(Y - P_S S - P_T T)$$

(Equation 6-4).

Maximizing L will also maximize V when the constraint inside the parenthesis is satisfied. The first-order conditions for a maximum obtain where

$$\frac{\partial L}{\partial S} = \frac{\partial V}{\partial S} - \lambda P_S = 0;$$

$$\frac{\partial L}{\partial T} = \frac{\partial V}{\partial T} - \lambda P_T = 0; \quad and$$

(Equation 6-5).

$$\frac{\partial L}{\partial \lambda} = 1 - P_S S - P_T T$$

Rearranging terms and dividing yields the optimal relationship between resources S and T:

$$\frac{\partial V / \partial S}{\partial V / \partial T} = \frac{P_S}{P_T}$$

(Equation 6-6).

That is, the ratio of each resource's marginal contribution to the total value of ecological resources protected will be equal to the ratio of the unit prices of protecting them. Since price and marginal cost are the same, we can express this relationship rather simply in terms of marginal benefits and costs:

$$\frac{MB_S}{MB_T} = \frac{MC_S}{MC_T}$$

(Equation 6-7).

In short, the ratio of the marginal benefits must equal the ratio of the marginal costs. This result holds regardless of the number of ecological resources involved.

References

Arrow KJ, Fisher AC. 1974. Environmental preservation, uncertainty, and irreversibility. *Q J Econ* 88(2):312-319.

Baumol WJ, Oates WE. 1975. The theory of environmental policy. Baltimore MD: Johns Hopkins University Press.

Ciriacy-Wantrup SV. 1952. Resource conservation: Economics and politics. Berkeley CA: University of California Press.

Coase RH. 1960. The problem of social cost. *J Law Econ* 2:1-40.

Cropper ML and Oates WE. 1992. Environmental economics: a survey. *J Econ Lit* 30:675-740.

Dasgupta P, Heal G. 1979. Economic theory and exhaustible resources. Cambridge MA: Cambridge University Press.

D'Arge RC, Kokigu KC. 1973. Economic growth and the environment. *Rev Econ Studies* 40(1):61-77.

Dales JH. 1968. Pollution, property and prices. Toronto, Canada: University of Toronto Press.

Devarajan S, Fisher AC. 1981. Hotelling's 'economics of exhaustible resources': 50 years later. *J Econ Lit* 19(1):65-73.

Fisher AC, Krutilla JV, Cicchetti CJ. 1972. The economics of environmental preservation: A theoretical and empirical analysis. *Amer Econ Rev* 62(4):605-619.

Fisher AC, Peterson FM. 1976. The environment in economics: A survey. *J Econ Lit* 14(1):1-33.

Gray LC. 1913. The economic possibilities of conservation. *Q J Econ* 27(3):497-519.

Hotelling H. 1931. The economics of exhaustible resources. *J Political Econ* 39(2):137-175.

Kotok EI. 1945. International policy on renewable natural resources. *Amer Econ Rev* 35(2):110-119.

Krutilla JV. 1967. Conservation reconsidered. *Amer Econ Rev* 57(4):777-786.

Krutilla JV, Fisher AC. 1975. The economics of natural environments. Baltimore MD: Johns Hopkins University Press.

Kryuchkov VV. 1993. Extreme anthropogenic loads and the northern ecosystem condition. *Ecol Appl* 3(4):622-630.

Pryde PR. 1983. The "decade of the environment" in the Soviet Union. *Sci* 220(4594):274-279.

Weitzman ML. 1992. On diversity. *Q J Econ* 107(2):157-183.

CHAPTER 7

Approaches to Ecological Risk-Based Decision-Making in a U.S. Federal Regulatory Context

Randall S. Wentsel, Maurice G. Zeeman, Anthony F. Maciorowski, Richard L. Orr, Stephen J. Ells

Introduction

Decision-makers must often utilize various risk management parameters (i.e. cost/benefit, engineering, societal issues, political issues, and risk assessment) in different ways and with different degrees of priority when evaluating environmental issues. Ecological and natural resource issues can be especially difficult to resolve because of the lack of regulatory bright lines in the determination of harm or change to an ecological system. However, with more difficult and complex environmental issues to consider, environmental policymakers have turned to a greater extent to risk assessment to assist them as part of the risk management process. However, risk assessment cannot be applied indiscriminately to all environmental issues. A relatively simple solution or treatment may be all that is required. In addition, public or political needs may require action based only on available information.

A recent White House Office of Science and Technology Policy (OSTP) report listed several pertinent questions concerning the utilization of risk assessment in environmental policy (OSTP 1995).

- When should a risk assessment be undertaken? When will it enhance policy decisions?
- How should a risk assessment be used and presented? How can risk assessment best be used to inform environmental policy and management decisions?
- What is the appropriate level of effort and precision to dedicate to risk assessment?
- As a means to ensure accountability and transparency, what is the most effective means to characterize and communicate information about risks, uncertainty in assessments, and limits to assessments?
- How might uncertainty be weighed into policy decisions?

Risk Management: Ecological Risk-Based Decision-Making. Ralph G. Stahl, Jr. et al., editors.
©2001 Society of Environmental Toxicology and Chemistry (SETAC). ISBN 1-880611-26-0

- What is the estimated incremental value of obtaining additional information through increased research, versus the incremental cost of delaying a decision?
- How should environmental justice and other social, cultural, and ecological concerns be integrated into framing a risk assessment and defining relevant data needs?

Policymakers who support the use of risk analysis believe that federal programs designed to protect public health, safety, and the environment should be carefully targeted to address the worst risks first and that the risk reduction achieved should be worth the cost. Politicians also hear complaints about unfunded federal mandates to state and local governments and the growing cost of compliance of environmental requirements to industries. Therefore, the balancing of risk reduction with other risk management considerations is important.

Opponents of this use of risk assessment argue that exclusive reliance on risk assessment and costs to define problems and evaluate solutions ignores other equally important facets of policy decisions, such as timeliness, fairness to all segments of society, and practicality. They also state that risk assessment and cost-benefit analysis undervalue environmental and health benefits, exaggerate costs, and focus on relatively small costs and risks spread throughout the entire population.

Uses of Risk Assessment

The federal government has several functions that, either directly or indirectly, act to reduce risks to public health, safety, and the environment. In this use, risk analysis is a tool for evaluating what is known about things that cannot be known with certainty, either due to effects of hazards that are unpredictable, lack of scientific understanding of data, uncertainty with models, and other extrapolations. Risk analyses produce estimates and they vary due to the quality of information used by the assessor. Risk assessors can only discuss the likelihood of various outcomes, the consequence of the outcome, the role of scientific judgment, and the present risks as statistical probabilities. Government agencies and departments evaluate risks, formally or informally, in most of their activities. Risk assessment is appropriate as an analytical tool to help identify problems, set regulatory priorities, compare effectiveness of risk management options, communicate to the public, and identify research needs. Since the purpose of environmental regulations is to protect human health and/or the environment, risk assessment, quantitatively or qualitatively, will estimate needed protection levels. Risk assessment is often involved in the generation of health or environmental criteria used in the regulations. Typically, risk assessment alone will not provide a hard and fast number for regulation. Uses of risk assessment include (CENR1995):

- defining problems and predicting risks,
- selecting risk avoidance or mitigation strategies and developing management programs,
- setting standards to protect human health or ecological health and evaluating ongoing risk reduction activities, and
- determining management and policy priorities.

Policy or nonscientific decisions in risk assessment include default assumptions and uncertainty analysis of scientific data. These decisions become problematic when two federal agencies evaluate the same data and calculate different levels of risk. Risk assessment techniques typically rely on multiple assumptions which may be untested, or even untestable. Risk assessment assumptions include the relevance of extrapolation of animal data, the choice of a dose-response model to extrapolate from high (experimental exposures) to the lower environmental doses, exposure pathways, and protection levels i.e. 95% of the effected population (Schierow 1995).

Actions Driven by Executive Orders

Since February 1981, when President Reagan issued Executive Order (EO) 12291, cost-benefit analyses have been required for major rules—i.e., rules likely to result in "[a]n annual effect on the economy of $100 million or more;" a "major increase in costs or prices for consumers, individual industries, federal, state, or local government agencies, or geographic regions." (Schierow 1993). However, on 30 September 1993, President Clinton issued EO 12866 which repealed EO 12291 while adding some specific requirements for regulatory developments. Now federal agencies were required to "consider the degree and nature of the risks posed by various substances or activities within its jurisdiction" and to conduct cost-benefit analysis for all "significant regulatory actions", including any substantive action expected to lead to promulgation of a "major rule" (as defined in EO 12291). This also applied to other rules that may adversely affect the economy, the environment, public health or safety, the state, local, or tribal governments, or the communities; create a serious inconsistency with an action taken or planned by another agency; alter the budgetary impact of entitlements, grants, user fees, loan programs, or the rights and obligations of recipients thereof; or raise novel legal or policy issues arising out of legal mandates, the President's priorities, or the principles for regulatory planning and review that are set out in the order and summarized below (EO 1993). Executive Order 12866 requires that regulations, including environmental and health regulations, be assessed for cost and benefit to the public. It includes assessing the nature of the risk, seeking views of state and local governments, using comparative risk assessment in regulatory decision making, and evaluating alternative approaches.

The White House, in 1996, released benefit-cost analysis guidance to support EO 12866 (RPR 1996). The document addresses three basic components for the economic analysis: 1) statement of need for the proposed action, 2) an examination of alternative approaches, and 3) an analysis of benefits and costs. Economic analysis refers to any systematic procedure to evaluate real or anticipated resource expenditures and losses (costs) relative to real or anticipated gains (benefits). Cost-benefit-risk analysis is the quantification and monetary valuation of the expenditures, gains, and losses, and the calculation of net benefits to society associated with the adoption of a particular regulation (or alternative management strategy) to address an environmental hazard.

Issues of debate in the guidance concern for example, monetizing of nonmonetary benefits. This would include values for ecosystems, future use of environmental resources, and discounting future health risks. Assessing uncertainty is also controversial because economists state that costs of a regulation are fairly certain, while benefits and risk calculations have a much larger uncertainty and should weigh less in the decision making process. Risk reduction is typically a part of the benefit. When benefits are calculated in monetary terms to allow cost-benefit-risk assessment, the potential for inaccuracies, value judgments, and lack of appropriate methods can arise. The various benefit techniques consist of calculating the dollar values of health effects, which include studies of how much people are willing to pay to avoid exposure to a hazard or particular adverse effect, or savings of direct costs, such as health care expenditures, salary loss for the duration of an illness, or the years of work lost to premature death. The intent is to estimate the gross monetary value of benefits to society and individuals. These numbers are then compared to costs of the regulation.

The U.S. Environmental Protection Agency's (USEPA's) Science Advisory Board (USEPA 1990) was critical of methods that assume that the future value of an ecological resource must be less than its present value. In the 1990 report, they stated that this policy inevitably leads to depletion of irreplaceable natural resources. Reliance on measures such as the public's willingness to pay exacerbate the problem because although the public may not care about wetlands, for example, they nonetheless are valuable now and in the future. They concluded that new techniques are needed to assess the real long-term value of ecosystems.

Environmental justice advocates put forward that certain subgroups may be burdened with a disproportionate share of environmental risks. In addition, other groups may gain a disproportionate share of the risk reduction, while taxpayers and consumers bear the cost of implementation and compliance. Instead of, or in addition to, weighing a regulation's total costs against total benefits, they want inequities to be described and avoided (Schierow 1995).

The Role of Federal Environmental Statutes

In the implementation of environmental statutes by the USEPA and other agencies, there are variations in how risk is addressed in the regulations. There is also variation between USEPA Offices in the parameters incorporated into a risk assessment. Both of these variables affect the determination of the appropriate use of risk assessment in implementing the regulations stemming from diverse environmental statutes. A uniform approach would enhance the usefulness of risk assessment as a regulatory tool.

The environmental statutes are written as either narrative or numerical directives. The USEPA implements narrative directives primarily through three different approaches: health based standards, technology based standards, and no unreasonable (balanced) risk standards. Table 7-1 presents examples of how risk is addressed in federal statutes. Health based standards regulate on the protection of human health or the environment without regard to technology or cost factors. The primary health-based standard was the Delaney Clause. In contrast, the Endangered Species Act (ESA) currently requires the protection of listed biota without regard to technology or cost factors. However, the ESA does not contemplate a zero-risk standard.

Technology-based standards require best practicable control technology, best available technology, and other controls for pollution reduction or treatment. Parameters evaluated include the determination of the effectiveness of methodologies on concentration reduction and often costs, rather than reduction of risk. For example, the Clean Water Act (CWA) requires technology-based controls to treat water pollution. Further, CWA requires effluent standards to provide an ample margin of safety, taking into account "the toxicity of the pollutant, its persistence, degradability, the usual or potential presence of the affected organisms in any waters, the importance of the affected organisms, the nature and extent of the effect of the toxic pollutant on such organisms, and the extent to which effective control is being or may be achieved under other regulatory" (33 U.S.C.1251; Sec.307.3001(b)(2)(A)). Yet, CWA does not instruct the USEPA on how it should balance these considerations relative to one another. Two sets of water quality criteria are used to protect aquatic life and to protect 'uman health. These water-quality criteria are based on risk assessment input. More recently, criteria to protect aquatic life have been set based on effluent bioassays or in-stream criteria. The criteria are the goal the technology-based standards seek to achieve. The use of risk assessment to support technology-based standards is an appropriate use of risk assessment. Many of the other statutes that use ecological risk assessment (ERA) i.e. Toxic Substances Control Act (TSCA), the Federal Insecticide, Fungicide and Rodenticide Act (FIFRA), and the Comprehensive Environmental Response, Compensation, and Liability Act (CERCLA) use a balanced risk-based approach.

Table 7-1 Approaches to risk in federal statutes (OSTP 1995)

Statute	Regulatory authority	Effects of concern	Approach to risk
Consumer Product Safety Act	Consumer products	"An unreasonable risk of injury"	Balance risks against product utility, cost, and availability
Occupational Safety and Health Act	Risks in the work place	"Material impairment of health or functional capacity": what is reasonably necessary or appropriate to provide safe and healthful employment?	Attain highest degree of health and safety protection: technical and economic feasibility
Federal Water Pollution Control Act (Clean Water Act [CWA])	Waste water discharges	"The toxicity of the pollutant, its persistence, degradability..."	Best available technology that is economically achievable
Clean Air Act Section 109	National ambient air quality standards	Protect public health	Set standards to provide ample margin of safety
Section 112	Emissions standards for hazardous air pollutants	Adverse effects to health and the environment	Reduce emissions using maximum achievable control technology, and later address "residual risk"
Section 202	Emissions standards for new motor vehicles	Unreasonable risk to health, welfare, or safety	Greatest degree of emission reduction achievable through technology available, taking into consideration cost, energy, and safety factors
Federal Insecticide Fungicide, and Rodenticide Act (FIFRA) Section 3	Pesticides	Unreasonable risks to health and the environment	Balance risks against economic benefits to pesticide users and society
Toxic Substances Control Act (TSCA) Section 6	Existing chemicals in commerce	Unreasonable risks to health and the environment	Balance risks against economic benefits, considering alternative technologies
Comprehensive Environmental Response, Compensation, and Liability Act (CERCLA) Section 313	Toxic release inventory	Hazards to human health or the environment	Reporting is based largely on hazard and quantity used
Section 9621	Hazardous waste site remediation	Persistence, toxicity, mobility, and propensity to bioaccumulate, short and long-term health effects	Protect human health and the environment in cost-effective manner
Safe Drinking Water Act Section 300g-1(b)	Drinking water	Known or anticipated adverse effects on human health	Set a goal (maximum contaminant level goal) with an adequate margin of safety, and define a maximum contaminant level as close as feasible to the goal

Ecological Risk-Based Decisions under Toxic Substances Control Act

The USEPA's Office of Pollution Prevention and Toxics (OPPT) conducts ERA and ecological risk management (ERM) for new chemical substances regulated under the TSCA. Under TSCA, manufacturers and importers of new industrial or general chemicals are required to submit a premanufacturer notification (PMN) to USEPA before they intend to begin manufacturing or importing. The OPPT has only 90 days to complete the risk assessment and with limited information reach a risk management decision on the chemical.

More than 70,000 chemicals are in commerce today and TSCA requires consideration of both new and existing industrial chemicals in the PMN program. In addition to the short review time allowed, three major difficulties are associated with evaluating PMNs: 1) confidential business information protection afforded by TSCA which restricts how information can be disseminated; 2) the fact that upwards of 2,200 new chemical notices are submitted to USEPA annually (Zeeman et al. 1995; Zeeman 1997); and 3) the limited information provided including chemical identity; molecular structure; trade name; production volume, use, and amount for each use; by-products and impurities; human exposure estimates; disposal methods; and any test data that the submitter may have. The manufacturer does not have to initiate any chemical property or ecological or human health testing before submitting a PMN. Only about 5% of the PMNs reviewed to date contain ecological effects data, and most of those data consist of acute toxicity tests performed on fish (Zeeman et al. 1993, 1995; Zeeman 1995).

Because of the paucity of data, there is a heavy reliance on the use of structure-activity relationships (SARs) to predict ecotoxic effects and exposure/fate characteristics (such as physical/chemical properties and biodegradation). Uncertainty (assessment) factors are used to adjust ecotoxicity values and compensate for a lack of definitive data when comparing effects concentrations with exposure levels. Given the constraints on the assessments, it is not possible to quantify the effects on the assessment endpoint: populations and communities of aquatic organisms and aquatic ecosystems. Nevertheless, the risk-assessment approach provides a useful way of applying scientific information to environmental decision-making.

The ERA for a new industrial chemical (Figure 7-1) has been in place for more than a decade (Zeeman and Gilford 1993). The OPPT's overall approach for new chemicals is to compare potential ecological effect concentrations that have been adjusted for uncertainty (i.e., concern concentrations) with potential exposure concentrations. The process often begins with estimating toxicity, adjusting these effect concentrations for uncertainty, and contrasting one or more of these concern concentrations (CC) with one or more predicted environmental concentrations from simple stream flow dilution models that typically result in reasonable worst-case exposure scenarios. If a risk is ascertained, more detailed analyses (e.g., refined exposure

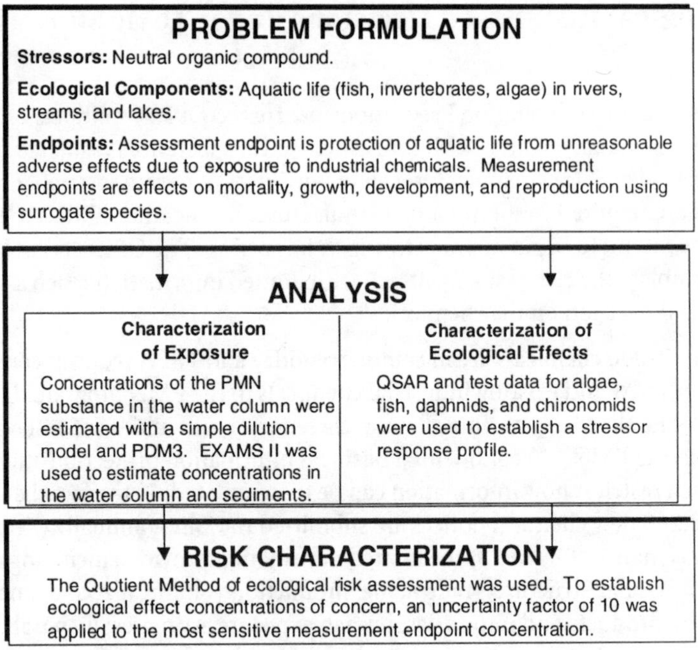

PROBLEM FORMULATION

Stressors: Neutral organic compound.

Ecological Components: Aquatic life (fish, invertebrates, algae) in rivers, streams, and lakes.

Endpoints: Assessment endpoint is protection of aquatic life from unreasonable adverse effects due to exposure to industrial chemicals. Measurement endpoints are effects on mortality, growth, development, and reproduction using surrogate species.

ANALYSIS

Characterization of Exposure	Characterization of Ecological Effects
Concentrations of the PMN substance in the water column were estimated with a simple dilution model and PDM3. EXAMS II was used to estimate concentrations in the water column and sediments.	QSAR and test data for algae, fish, daphnids, and chironomids were used to establish a stressor response profile.

RISK CHARACTERIZATION

The Quotient Method of ecological risk assessment was used. To establish ecological effect concentrations of concern, an uncertainty factor of 10 was applied to the most sensitive measurement endpoint concentration.

Figure 7-1 Structure of assessment for effects of a premanufacture notification substance

scenarios, such as EXAMS II, and/or ecotoxicity testing results) are sought to inform decision making (Table 7-2 and Figure 7-2). As noted previously, there is a heavy reliance on the use of quantitative structure-activity relationships to predict ecotoxic effects to develop a stressor-response profile or an ecotoxicity profile (Zeeman et al. 1993, 1995).

Table 7-2 An example of five risk characterization iterations

Iteration	Estimates/assumptions	Uncertainty
1	Fish are the most sensitive species. CC set at 1 µg/L. PMN substance mixes instantaneously in water. No losses.	Worst-case analysis.
2	Actual test data for daphnids still indicate a CC of 1 µg/L. Determine how often this concentration is exceeded using refined exposures (PDM3).	Worst-case analysis. Other species may be more sensitive
3	Estimate risk to benthic organisms using daphnid ChV and mitigation by organic matter. EXAMS II used to further refine estimated concentrations.	Generic production sites. Actual data for benthic organisms not available.
4	Site-specific data obtained on use and disposal. EXAMS II rerun with new data.	Estimated toxicity for benthic invertebrates.
5	Actual test data for benthic organisms obtained.	Best estimates for identified sites. May not hold for other sites or uses.

CC=concern concentrations

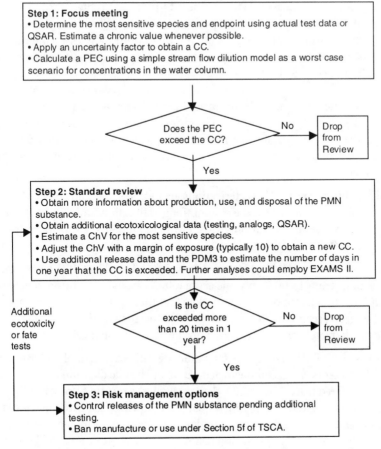

Step 1: Focus meeting
• Determine the most sensitive species and endpoint using actual test data or QSAR. Estimate a chronic value whenever possible.
• Apply an uncertainty factor to obtain a CC.
• Calculate a PEC using a simple stream flow dilution model as a worst case scenario for concentrations in the water column.

Does the PEC exceed the CC?   No → Drop from Review

Yes

Step 2: Standard review
• Obtain more information about production, use, and disposal of the PMN substance.
• Obtain additional ecotoxicological data (testing, analogs, QSAR).
• Estimate a ChV for the most sensitive species.
• Adjust the ChV with a margin of exposure (typically 10) to obtain a new CC.
• Use additional release data and the PDM3 to estimate the number of days in one year that the CC is exceeded. Further analyses could employ EXAMS II.

Additional ecotoxicity or fate tests

Is the CC exceeded more than 20 times in 1 year?   No → Drop from Review

Yes

Step 3: Risk management options
• Control releases of the PMN substance pending additional testing.
• Ban manufacture or use under Section 5f of TSCA.

Figure 7-2 Flow chart and decision criteria for the ecological risk assessment of a premanufacture notification substance

Figure 7-2 does not include all likely risk management options. In addition to obtaining additional exposure and ecological effects information, risk management options can include a variety of regulatory enforcement actions, such as requiring pretreatment or even banning discharges to water. In any event, OPPT risk assessors must ascertain that a risk exists before OPPT risk managers need to exercise these risk management options.

The approach taken in PMN evaluation has the following strengths: 1) it relates measurement endpoints to an assessment endpoint, 2) it demonstrates that ERAs can be conducted with minimal toxicity data and exposure data, 3) it demonstrates the usefulness of SARs in establishing a toxicity or stressor-response profile, and 4) it demonstrates that regulatory decisions can be made quickly using only the best data available at the time.

For new industrial chemical evaluations, the ERA process (problem formulation, analysis, and risk characterization) is applied in a tiered fashion. The initial planning and problem formulation stage is quite similar for most assessments because the assessments are usually not site specific and similar models and endpoints are used for different chemicals. Assessment endpoints and measures of effect (measurement endpoints) are identified, and the analysis and risk-characterization phases are conducted sequentially using additional data and fewer worst case assumptions with each successive tier. The overall approach is to compare potential ecological effect concentrations that have been adjusted for uncertainty with potential exposure concentrations. If a risk is ascertained, more detailed analyses are performed.

Generally, a number of risk characterizations are performed for any one PMN. Table 7-2 provides a brief summary of the assumptions, estimations, and types of uncertainty for each of five such iterations. Most PMNs (about 80%) are dropped from further review after the initial iteration.

The assessment (or uncertainty) factors developed and used by OPPT (presented in Table 7-3) range from 1,000 if only one acute value is available to 1 from a field study (e.g., pond) or from a microcosm study. Note that these assessment factors are designed to decrease in magnitude as more definitive toxicity data are made available to adequately assess the hazard profile of a new chemical.

If the results of the initial risk characterization identified a potential unreasonable risk, OPPT assessors can recommend acute and/or chronic daphnid tests to complete the tiered tests. The USEPA also can request a test that simulates the effectiveness of a wastewater treatment plant in removing the PMN substance from the waste stream. A second iteration could occur if it was determined that a potentially unreasonable risk to aquatic organisms still existed. Additional toxicity tests to clarify unanswered questions may be required along with a refined exposure evaluation with the use of models (Burns 1989). As is evident from the new chemical assessment example, there can be up to five iterations in characterizing the risk of this chemical to organisms in the environment (Table 7-2). However, about 80% of the PMNs are resolved at the first tier. The risk management decisions made play a key role in deciding the next steps for each of these iterations. This illustrates how

Table 7-3 Office of Pollution Prevention and Toxics assessment factors used in setting "concern levels" for new chemicals

Available data on chemical or analogue	Assessment factor
Limited (e.g., only one acute LC50 via SAR/QSAR)	1000
Base Set acute toxicity (e.g., fish and daphnia LC50s and algal EC50)	100
Chronic toxicity MATCs[1]	10
Field test data for chemicals	1

[1]MATC = maximum acceptable concentration
adapted from Nabholz 1991; Zeeman and Gilford 1993; USEPA 1984

an efficient and pragmatic ERA process can assist in eliciting reasonable ERM decisions. These ERM decisions helped to develop the kinds of information needed to perform an adequate ERA and to come to closure with the regulatory actions determined to address and/or mitigate the ecological risks expected from allowing the use of this chemical.

Ecological Risk-Based Decision-Making under the Federal Insecticide, Fungicide and Rodenticide Act

The USEPA has the regulatory authority under FIFRA to register pesticides to ensure that there are no unreasonable adverse effects to human health or the environment, taking into account the economic, social, and environmental costs and benefits of the pesticide use. This is a balanced risk-based approach where FIFRA is a cost-benefit statute, and an unreasonable adverse effect on the environment is a regulatory determination that must account for scientific as well as economic, social, and governmental cost and benefit factors.

In a given year, the USEPA's Office of Pesticide Programs (OPP) reviews about 5,000 registration submissions. To ensure compliance with current scientific and regulatory standards, FIFRA also requires the review and reregistration of existing pesticides. During reregistration, registrants may delete pesticide uses or voluntarily withdraw products or uses. In addition, the USEPA has the authority to cancel registrations for pesticide products that do not meet the requirements for reregistration. Since 1988, the registered products subject to reregistration have declined from approximately 50,000 to about 20,000. Problems that arise during the use of a particular pesticide, following registration or reregistration, may be investigated under the special review process. Special review consists of scientific and legal analysis before a major regulatory decision is made on a registered pesticide.

The integrated decision-making process involves the following three interactive phases:

1. Risk assessment is a science-based activity that consists of hazard characterization and exposure characterization and ultimately integrates the two into a risk characterization.

2. Risk mitigation involves remediation or mitigation measures to reduce or eliminate source contamination and adverse environmental impacts.

3. Risk management is a policy-based activity that defines risk-assessment questions and endpoints to protect human health and the environment. It takes the scientific risk assessment, incorporates social, economic, political, and legal factors that impinge on or influence the final decision, and selects regulatory actions.

In the registration and reregistration process, decisions are based in part on the evaluation, synthesis, and integration of pesticide studies conducted by registrants and submitted to the Agency. The environmental fate and transport and ecological effects studies (40 CFR 158.130; 40 CFR 158.145) are evaluated by the USEPA scientists and subsequently used in an ERA (Urban and Cook 1986; Touart 1988; Fite et al. 1988; SETAC 1994; Touart 1995; Touart and Maciorowski 1997) that is consistent with the generalized ERA framework (USEPA 1991).

The scope and complexity of any specific pesticide risk assessment will vary with the specific chemical and use. The ERA is generally tiered, progressing from simple risk quotients derived from laboratory fate, transport, and toxicity data in early tiers to a weight-of-evidence approach in later tiers (Tables 7-4 and 7-5).

Exposure analysis may consist of a preliminary or comprehensive fate and transport assessment (Table 7-4) based on registrant-submitted data. The exposure analysis provides exposure profiles and estimated environmental concentrations (EEC). Estimated environmental concentrations may be derived from four estimation procedures ranging from simple to complex. Preliminary exposure analysis includes simple laboratory tests and models to provide an initial fate profile for a pesticide (hydrolysis and photolysis in soil and water, aerobic and anaerobic soil metabolism, and mobility).

Fate and transport assessment provides a comprehensive profile of the chemical (persistence, mobility, leachability, binding capacity, degradates) and may include field dissipation studies, published literature, other field monitoring data, ground-water studies, and modeled surface water estimates. The ecological-effects analysis (Table 7-4) is also tiered. Tier I provides an acute toxicity profile for birds, fishes, mammals, and invertebrates. Tier II provides a subchronic and chronic toxicity (no-observed-effects concentration or NOEC) profile and bioaccumulation potential for the same test species. Depending on the hazard and exposure characteristics of a particular pesticide and use pattern, Tier II analyses may be conducted for all

Table 7-4 Generalized exposure analysis and assessment methods and procedures used in prospective ecological risk screens of pesticides[1]

Level	EEC estimation procedures
1	A direct-application, high-exposure model designed to estimate direct exposure to a nonflowing, shallow-water (<15 cm) system
2	Adds simple drift or runoff exposure variables such as drainage basin size, surface area of receiving water, average depth, pesticide solubility, surface runoff, or spray drift loss, which attenuate the Level 1 direct application model estimate
3	Computer runoff and aquatic exposure simulation models. A loading model (SWRBB-WQ[2], PRZM[3], etc.) is used to estimate field losses of pesticide associated with surface runoff and erosion; the model then serves as input to a partitioning model (EXAMS II[4]) to estimate sorbed and dissolved residue concentrations. Simulations are based on either reference environment scenarios or environmental scenarios derived from typical pesticide use circumstances
4	Stochastic modeling where EECs are expressed as exceedance probabilities for the environment, field, and cropping conditions

[1]For additional details regarding environmental fate data requirements, see 40 CFR § 158.130, SETAC 1994, Touart1995
[2]Simulator for water resources in rural basins—water quality
[3]Pesticide root zone model
[4]Exposure analysis modeling system

Table 7-5a Tiered effects analysis used in prospective risk screens of pesticides

Tier I	provides acute toxicity values and dose-response information (mammalian and avian acute oral LD50; avian dietary LC50; seedling emergence and vegetative vigor EC25; honeybee acute contact LD50and additional wild mammal, estuarine, and plant tests depending on pesticide use category).
Tier II	provides subchronic and chronic toxicity values (NOEC) including avian reproduction studies; special avian or mammal studies; fish early life stage studies; invertebrate life cycle studies; and a fish bioaccumulation factor.
Tier III	provides refined NOEC estimates for chronic toxicity that may include a fish full life cycle test, aquatic organism accumulation, or food chain transfer tests.
The quotient method	provides a set of acute and chronic RQ for fishes, birds, invertebrates, plants, and endangered species. The RQs are calculated by dividing exposure (EEC) by hazard (LD50 or LC50 or NOEC).
Tier IV	allows registrants to rebut a presumption of risk derived from laboratory studies by performing field or simulated field studies, including qualitative terrestrial field studies, farm pond studies, mesocosm studies, or other special studies.

Table 7-5b Regulatory risk criteria that risk quotients are compared to

Presumption of acceptable risk	Presumption of risk that may be mitigated by restricted use	Presumption of unacceptable risk	
		Non-endangered species	Endangeı
Acute toxicity EEC<0.1 LC50	0.1 LC50 EEC 0.1 LC50	EEC 0.50 LC50	EEC 0.05 EC25 0.1ı
Chronic toxicity EEC chronic NOEC	N/A	EEC > NOEC	

RQ=risk quotient
EEC=estimated environmental concentration
NOEC=no-observed-effects concentration

representative taxa, or may focus on either aquatic or terrestrial species. When warranted, Tier III effects analysis is used to refine NOEC and bioaccumulation estimates.

Following exposure and effects analysis, ecological risk is estimated as a function of ecotoxicological effects and environmental exposure using the quotient method (Table 7-4). A number of risk quotients are calculated and compared with regulatory-risk criteria (e.g., presumption of acceptable risk, presumption of unacceptable risk, etc.). Traditionally, if regulatory criteria are exceeded, a high-risk potential is assumed to exist for the pesticide-use combination. If a registrant wishes to refute a presumption-of-risk finding, a Tier IV effects analysis, consisting of field studies, simulated field studies, or other special studies, may be conducted (Fite et al. 1988; Touart 1988).

The application of ERA in pesticide regulatory decisions is subject to practical constraints imposed by law, regulatory policy, and precedent. When a pesticide undergoes evaluation for registration or reregistration, the scientific experts review and evaluate the data available in a comprehensive manner to ensure it meets the standards established for carrying out risk assessments. This information, along with the hazards of the pesticide as determined in the required studies and available incident data, is used to determine what level of concern exists in each of several compartments in the environment. If a level of concern is unacceptable, then risk mitigation/verification procedures are initiated. In the risk management decision-process, the risk assessor may be asked to analyze or judge the effect of proposed risk mitigation on the original risk assessment. This does not change the original risk assessment, which serves as a baseline estimate, but the analysis may begin here as to whether management actions such as mitigation will reduce risk to acceptable levels.

In the registration and reregistration processes, a conclusion that an unacceptable risk will result from the proposed or registered use of pesticides triggers an interactive process of identifying appropriate risk reduction measures. Once OPP and the registrant have concluded their work on appropriate risk mitigation steps, negotiations between OPP and the registrant occur in the form of risk management actions including changes or restrictions for specific uses or label changes. As this process begins, data to support the effectiveness of the mitigation steps will be nonexistent or limited in scope. To ensure the effectiveness of the mitigation steps, the agency may require some verification data. Once mitigation measures have been identified and implemented, post-registration or post-reregistration monitoring may be required to verify the efficacy of the risk mitigation measures. Quantifiable verification of effectiveness of the mitigation may take several years. The verification data would then be reviewed to evaluate the effectiveness of the mitigation measures. In this sense, the approach is very similar to the framework discussed in Chapter 2.

Risk managers are the ultimate users of pesticide ERAs. Presented with a scientific evaluation of risk, the risk manager may want additional information or studies, or the risk manager may need to act on the information in hand regardless of its scientific strengths or shortcomings. Rather than refine the ERA, a risk manager may opt to impose mitigation to reduce the risk, even in the face of uncertainty with hope that the mitigation will be effective. When such situations occur, it is important for risk assessors to clearly and succinctly summarize risk, uncertainties, and options for the benefit of risk managers, stakeholders, and the public. However, risk managers must also consider other factors in the decision-making process, which is often subject to decidedly unscientific constraints within existing law, policy, and socioeconomic and political realities. Risk assessors must be aware of ERM needs in the problem formulation stage to ensure that the assessment endpoints and resolving power that the decision maker requires are understood. Ideally, this should be agreed to during a formal a priori problem-formulation step in the assessment process. Routine problem formulation that engages both risk assessors and risk managers is increasing, but has not been commonly practiced in the past. This has sometimes led to differing expectations between risk assessors and risk managers regarding the objectives, scope, and application of an ERA. The importance of promoting formal problem formulation cannot be overstated.

Ecological Risk-Based Decision-Making under Comprehensive Environmental Response, Compensation, and Liability Act

Ecological Risk Assessment under CERCLA are retrospective evaluations of the effects of contamination in a given area. They provide baseline information on whether a cleanup should be considered for ecological reasons, and risk assessments are used in the evaluation of remedial alternatives.

Risk managers are required to protect human health and the environment and to comply with applicable, relevant, and appropriate requirements. They also balance the risk and proposed mitigation methods with various economic, societal, technical, and political concerns discussed in this chapter.

The primary purpose of performing an ERA at Superfund sites is to determine if releases or potential releases of hazardous substances from the site have resulted in or are likely to result in unacceptable adverse effects on ecological receptors. The goal of Superfund response actions is to prevent effects from occurring or, if effects have occurred, to implement a remedy that will provide adequate protection of the ecological resources in a cost-effective manner that also meets any appropriate federal or state laws. Ecological risk assessments should be designed to determine a threshold media concentration that will provide adequate protection of important ecological resources. This requires substantial discussion between the risk assessor and the risk manager before sampling to make sure the information needed to make these decisions is collected.

Ecological risk assessment data are used for
- characterizing baseline risk to determine whether a cleanup should be considered,
- deriving specific contaminant concentrations that provide adequate protection from unacceptable risks,
- evaluating the remedial alternatives for potential effectiveness and potential risks, and
- providing baseline information that can be followed with monitoring to document that the remedy is effective at reducing risk.

The performance and use of ERA in USEPA Records of Decision in 1995 is significant (Table 7-6). Ecological risk assessments are performed at 60% of the sites and the remedial activity at a site is based in part on ecological data about 46% of the time.

Table 7-6 Use of ecological data in Records of Decision (ROD) in 1995

Explanation	Total Number	Percentage of total RODs
Number or RODs with ecological risk assessments	113	60
Number or RODs where remedial action based at least partially on ecological risk	52	46
Number of RODs where population/community study performed	24	21
Number of RODs where modeling performed	25	22
Number of RODs where literature values used	50	44
Number of RODs where ambient water criteria used	19	17
Number of RODs where NOAA[1] sediment values used	11	10
Number of RODs where site-specific toxicity tests performed	14	12
Number of RODs where tissue sampling performed	10	9

[1]NOAA-National Oceanic and Atmoshperic Administration

Under CERCLA, whenever there is a release or a substantial threat of a release of a hazardous substance into the environment, USEPA is authorized to take whatever action is deemed appropriate to protect the environment, as long as the action is consistent with the National Contingency Plan (NCP). The USEPA uses the information from the risk assessment in its decision-making process. The NCP requires that a baseline risk assessment be conducted by the lead agency during the remedial investigation/feasibility study in order to characterize the current and potential threats to human health and the environment (Section 300.430[d][4]). Any remedy selected by USEPA must be protective of the environment and human health, comply with any enforceable federal or state standards or criteria, and balance a number of criteria including 1) long-term effectiveness and permanence of the response; 2) reduction of toxicity, mobility, or volume of the waste through treatment; 3) short-term effectiveness; 4) implementability; and 5) cost. Two modifying criteria also must be considered: state acceptance and community acceptance.

The Superfund program currently has few written policies or guidances that explicitly explain how to make reasoned ERM decisions, even though a draft framework has been developed (Luftig 1998). Often the decision-maker must rely on the guidance given in the NCP. Unlike human-health-risk assessments, which have quantifiable risk goals that define levels of acceptable risk to one species (e.g., to reduce human cancer risks to levels below 1 in 10,000), quantifiable ERM goals have not been established. The NCP states only, Alternatives shall be assessed to determine whether they can adequately protect human health and the environment, in both the short- and long-term, from unacceptable risks posed by hazardous substances (Section 300.430[e][9][iii][A]). This lack of a simple and easily articulated cleanup goal makes the selection of an appropriate remedy that is protective of the environment and meets the other eight NCP criteria problematic.

Superfund decision makers also have to consider nine criteria when selecting a response action that is appropriate for the site:

1. overall protection of human health and the environment;

2. compliance with applicable, relevant, and appropriate requirements (ARARs);
3. long-term effectiveness and permanence;
4. reduction of toxicity, mobility, or volume through treatment;
5. short-term effectiveness;
6. implementability;
7. cost;
8. state acceptance;
9. community acceptance.

The first two criteria are thresholds that must be met at every site (though the ARARs can be waived under certain circumstances), the next five are balancing criteria, and the last two are modifying criteria. The three criteria usually most important to the ecological risk manager are protection, long-term effectiveness, and short-term effectiveness.

Risk managers typically address three key questions at every Superfund site: 1) do site releases present an unacceptable risk to important ecological resources? 2) if the answer is yes, should the site be actively cleaned up or will the remedy do more damage (and thus not provide short-term protectiveness)? and 3) if cleanup is warranted, how do you select a cost-effective response and cleanup levels that provide adequate protection?

The ecological risk assessors consider the results from a battery of toxicity tests, field studies, and food-chain models to determine whether or not the observed or predicted adverse effects are unacceptable. The same studies are often used to select chemical-specific cleanup levels that are believed to be protective at that site.

Whether or not to clean up a site is often the most difficult risk-based decision to make. Even though an ERA may demonstrate that unacceptable ecological effects have occurred or are expected to occur in the near future, removal or in situ treatment of the contamination may do more ecological damage (often due to widespread physical destruction of habitat) than leaving it in place. In evaluating remedial alternatives, the NCP highlights the importance of considering the long-term and short-term impacts of the various alternatives in determining which alternatives adequately protect human health and the environment. A remedy that does significant short-term ecological damage often would not be considered to meet the NCP threshold criteria of protective.

Assuming remediation is technically practicable and not cost prohibitive, risk managers consider the long- and short-term ecological impacts of active remediation versus natural attenuation of the contaminants. The evaluation of ecological impacts from implementing remedial alternatives is part of the ERA process and should be discussed in a feasibility study. At sites with contaminants that degrade or with sediment contaminants that will become unavailable because of natural deposition of uncontaminated sediment over them, preventing additional releases may be the most appropriate remedy.

Before making a response decision, the risk manager, in consultation with an ecological risk assessor, often considers many of the following factors:

- the magnitude of the observed or expected impacts of site releases on the affected ecosystem component (e.g., fish population, benthic community),
- the likelihood that these impacts will occur or continue,
- the size and functional value of the impacted area in relation to the larger ecosystem,
- whether or not the impacted area is a highly sensitive or ecologically unique environment,
- the recovery potential of the impacted ecosystem and expected persistence of the chemicals of concern under the site conditions,
- short-term and long-term impacts of the remedial alternatives on the site habitat and larger ecosystem,
- effectiveness of the remedy (whether there are other continuing, nearby, non-Superfund releases or other types of stressors that will continue to adversely impact the ecosystem after the cleanup is implemented),
- community opinion on the value of the affected portion of the ecosystem and of the natural resources affected, and
- whether or not there will be any remaining residual risks that may need to be addressed by a natural resource trustee.

It is the responsibility of the risk manager, in consultation with the risk assessor, to select a remedy and ensure cleanup levels for the site that are reasonable. This decision can be made only after a thorough consideration of all nine criteria described in the NCP. Because of the high complexity of ecosystems and the large number of species potentially affected at every site, there will usually be a relatively high degree of uncertainty concerning the levels deemed to be protective—are they too high or too low? At these sites, monitoring of the affected ecological receptors should be performed after the remedy has been implemented in order to determine if recovery is occurring in a reasonable time frame and whether or not an additional response action is warranted.

Summary

Although each environmental statute approaches the problem of controlling risk from a different vantage point and authorization, they have two major issues of concern in the development of ecological risk assessment in the federal government. One is the effective communication of the science to risk managers and stakeholders, and the second is the utilization of more sophisticated tools to improve accuracy and to reduce uncertainty in the calculation of the likelihood and consequences of a regulated action. The topics listed in this chapter are all moving toward more effective communication to interested parties. The utilization of tiered approaches, the development of mitigation options, and the active coordination with risk managers in the problem formulation phase are all activities that increase effective communication. The consideration of more complex ecological systems for regulatory evaluation has required the application of various methods to determine the risk to these systems. These techniques include probabilistic methods to present distributions for exposure and effects and the utilization of distributional analysis in the presentation of risks to managers, the use of population models to assess stressor impacts, the application of geographic information system (GIS) techniques to assist in the evaluation of impacts at the watershed or regional level, and various statistical and modeling techniques to evaluate data. The utilization of these tools will hopefully enhance the determination of risks to ecological systems and the communication of these risks to interested parties.

The underlying principles behind risk reduction and integrated decision making are detailed in the strategic initiatives and guiding principles recently released by the USEPA (1994) and include ecosystem protection, pollution prevention, strong science and data, partnerships, and environmental accountability. Current policies are directed toward greater participation in environmental problem solving and decision making, including parties affected by the decision (regulated community, user groups, environmental interest groups, general public, and scientists). In the integrated decision-making process, risk assessors and risk managers, as well as other professionals (e.g., lawyers, economists, agronomists, etc.), interact and communicate during the development, interpretation, and final application of ERA. Although all involved parties should be aware of one another's spheres of expertise, influence, and control, this is not always the case. Because this is a relatively recent and rapidly evolving process, roles, responsibilities, information needs, and process boundary points are rarely well defined or well understood. Therefore, improved understanding of the different perspectives of risk assessors and risk managers is crucial to the ultimate success of integrated decision-making processes.

Until the overall integrated decision-making process is better defined and understood by both risk assessors and risk managers, there undoubtedly will be some controversy regarding the application of ERA in regulatory operations. However, recognizing and understanding that risk assessors and risk managers have different roles and responsibilities should go a long way toward improving the decision process.

References

Burns LA. 1989. Exposure analysis modeling system: User's guide for EXAMS II version 2.94. Athens GA: USEPA, Environmental Research Laboratory.

{CENR] Committee on Environmental and Natural Resources. 1995. Draft research strategy and implementation plan: Risk assessment research in the Federal Government. Washington DC: National Science and Technology Council.

[EO] Executive Order. EO12866 of 30 September 1993. Federal Register. 58(190):51735-51744.

Fite EC, Turner LW, Cook NJ, Stunkard C. 1988. Guidance document for conducting terrestrial field studies. Hazard Evaluation Division Technical Guidance Document. Washington DC: USEPA, OPP. EPA-540-09-88-109.

Luftig S. 1998. Draft ecological risk management principles for Superfund sites. OSWER Directive. 9285:7-28.

Nabholz JV, Clements RG, Zeeman, MG. 1993. Validation of structure activity relationships used by EPA's Office of Pollution Prevention and Toxics for the environmental hazard assessment of industrial chemicals. In: Gorsuch JW, Dwyer FJ, Ingersoll CG, editors. Environmental toxicology and risk assessment. Philadelphia PA: ASTM. STP 1216. p 571-590.

[OSTP] Office of Science and Technology Policy. 1995. Science, risk, and public policy. Washington DC: Executive Office of the President.

Portney P. 1990. The evolution of federal regulation. In: Portney P , editor. Public policies for environmental protection. Washington DC: Resources for the Future. p 7-26.

[RPR] Risk Policy Report. 1996. White house issues benefit-cost guidance, ending months of debate. Washington DC: Special Report.

Schierow L. 1993. Environmental risk and public policy, part I: Cost-benefit-risk analysis of environmental regulations. Washington DC: The Library of Congress, Congressional Research Service.

Schierow L. 1995. The role of risk analysis and risk management in environmental protection. Washington DC. The Library of Congress, Congressional Research Service. IB94036.

[SETAC] Society of Environmental Toxicology and Chemistry. 1994. Final report: Aquatic risk assessment and mitigation dialogue group. Pensacola FL: SETAC. 220 p.

Touart LW. 1988. Aquatic mesocosm tests to support pesticide registration. Hazard Evaluation Division Technical Guidance Document. Washington DC: USEPA, OPP. EPA-540-09-88-035.

Touart LW. 1995. The federal insecticide, fungicide and rodenticide act. In: Rand GM, editor. Fundamentals of aquatic toxicology. Washington DC: Taylor and Francis. p 657-668.

Touart LW, Maciorowski AF. 1997. Information needs for pesticide registration in the United States. *Ecol Appl* 74:1086-1093.

Touart LW, Maciorowski AF. 1997. Information needs for pesticide registration in the United States. *Ecol Appl* 74:1086-1093.

Urban DJ, Cook JN. 1986. Hazard evaluation division standard evaluation procedure. Washington DC: USEPA, OPP. EPA-540-19-83-001.

[USEPA] U.S. Environmental Protection Agency. 1990. Reducing risk: Setting priorities and strategies for environmental protection. Washington DC: USEPA Science Advisory Board. SAB-EC-90-021.

[USEPA] U.S. Environmental Protection Agency. 1994. The new generation of environmental protection. Washington DC: USEPA. EPA-200-B-94-002.

[USEPA] U.S. Environmental Protection Agency. 1998. Guidelines for ecological risk assessment. Federal Register 63(93):26846-26924.

Zeeman M. 1995 Ecotoxicity testing and estimation methods developed under section 5 of the Toxic Substances Control Act (TSCA). In: Rand G, editor. Fundamentals of aquatic toxicology: Effects, environmental fate, and risk assessment. 2nd edition. Washington DC: Taylor and Francis. p 703-715.

Zeeman M. 1997 Aquatic toxicology and ecological risk assessment: USEPA/OPPT perspective and OECD interactions. In: Zelikoff JT, Lynch J, Schepers J, editors. Ecotoxicology: Responses, biomarkers, and risk assessment. Fair Haven NJ: OECD. OECD [SOS Publications. p 89-108.

Zeeman M, Gilford J. 1993. Ecological hazard evaluation and risk assessment under EPA's Toxic Substances Control Act (TSCA): An introduction. In: Landis WG, Hughes JS, Lewis MA, editors. Environmental toxicology and risk assessment. Philadelphia PA: ASTM. STP 1179. p 7-21.

Zeeman M, Nabholz JV, Clements RG. 1993. The development of SAR/QSAR for the use under EPA's Toxic Substances Control Act (TSCA): An introduction. In: Gorsuch JW, Dwyer FJ, Ingersoll CG, editors. Environmental toxicology and risk assessment. Volume 2. Philadelphia PA: ASTM. STP 1216. p 523-539.

Zeeman M, Auer CM, Clements RG. 1995. U.S. EPA regulatory perspectives on the use of QSAR for new and existing chemical evaluations. *SAR QSAR Environ Res* 3(3):179-202.

CHAPTER 8

Adaptive Regulation of Waterfowl Hunting in the U.S.

Fred A. Johnson

Introduction

The harvest of renewable natural resources is predicated on the theory of density-dependent population growth (Hilborn et al. 1995). This theory predicts a negative relationship between the intrinsic rate of population growth and population density (i.e., number of individuals per unit of limiting resource) due to intraspecific competition for resources. In a relatively stable environment, unharvested populations tend to settle around an equilibrium where births equal deaths. Populations respond to harvest losses by increasing reproductive output or through decreased natural mortality. Population size eventually settles around a new equilibrium and the harvest, if not too heavy, can be sustained without destroying the breeding stock. Resource managers typically attempt to maximize the sustainable harvest by driving population density to a level that maximizes the intrinsic rate of population growth (Beddington and May 1977).

Although the theoretical basis for harvesting renewable resources is fairly straightforward, the practice of harvest management has had its share of difficulties. History is replete with cases where uncontrolled variation in harvests or the environment, naive assumptions about system response, and management policies with short time horizons have led to resource collapse (Ludwig et al. 1993). To be successful, sustainable harvesting depends on an ability to effectively regulate the size of the harvest, a sound understanding of the biological system and its density-dependent responses, and management objectives that are congruent with the renewal capacity of the resource. Even with a firm commitment to long-term resource conservation, harvest managers always will be burdened by complex, dynamic systems that are only partially observable and by management controls that are indirect and limited. It is for these reasons that a coherent framework for managing ecological risk is necessary.

Harvest-management decisions involve three fundamental components: 1) unambiguous objectives; 2) alternative harvest actions, including any constraints on those actions; and 3) predicted consequences of those actions in terms that are relevant to the stated management objectives. The consequences of harvest actions cannot be predicted with certainty, and the associated risk is what makes management

Risk Management: Ecological Risk-Based Decision-Making. Ralph G. Stahl, Jr. et al., editors.
©2001 Society of Environmental Toxicology and Chemistry (SETAC). ISBN 1-880611-26-0

decisions difficult. Risk is defined as the probability of a management outcome, where the probability can be assessed reliably from past experience with the resource or with a similar biological system. Risk differs from true uncertainty in which past experience provides no guide for the future (Costanza and Cornwell 1992). In keeping with the definitions in this book, ecological risk assessment involves associating empirical probabilities of possible system responses with alternative management actions. Ecological risk management (ERM) is the process of using management objectives to value those (probabilistic) responses so that a preferred management action can be identified.

The purpose here is to describe the process of risk assessment and management used to establish waterfowl hunting regulations in the U.S. This chapter first provides information about the regulations-setting process and about the biological monitoring and assessment programs that provide the basis for decision making. This chapter then provides a description of the conceptual framework and key features of waterfowl harvest management. Finally, an example of this framework is provided as it is applied to the management of mallard harvests.

Background

Federal regulations governing the sport hunting of waterfowl in the U.S. have significant biological and socioeconomic impacts. Each year, roughly 13 million waterfowl, principally mallards (*Anas platyrhynchos*), teal (*A. crecca, A. discors*), wood ducks (*Aix sponsa*), and Canada geese (*Branta canadensis*) are harvested by about 1.5 million sport hunters (USFWS 1988). In some cases, sport harvests represent up to 25 percent of the post-breeding population size (Anderson 1975). The impact of hunting activity on the economy also is significant. Waterfowl hunters in the U.S. spend over $500 million in pursuit of their sport, and the total economic output is estimated at $1.6 billion annually (Teisl and Southwick 1995).

The U.S. government's authority for establishing waterfowl hunting regulations is derived from treaties for the protection of migratory birds signed with Great Britain (for Canada in 1916), Mexico (1936), Japan (1972), and the Soviet Union (1978) (USFWS 1975). These treaties prohibit all take of migratory birds from 10 March to 1 September each year and provide for hunting seasons not to exceed 3 months. Each year, the U.S. Fish and Wildlife Service (USFWS) solicits proposals for hunting seasons from interested parties, and after extensive public deliberations, establishes guidelines within which states select their hunting seasons. States may be more restrictive, but not more liberal, than federal guidelines allow. Hunting regulations typically specify season dates, daily bag limits, shooting hours, and legal methods of take.

Waterfowl hunting regulations have worked reasonably well, as evidenced by the levels of hunting opportunity and harvest that have been maintained for at least 30 years. This record of success is notable given that natural resources often are overexploited to the point of economic extinction (Ludwig et al. 1993). This is not to say, however, that the process of setting waterfowl hunting regulations has been without problems. The process often is plagued by controversy, contentiousness, and, on occasion, court challenges and congressional intervention (Feierabend 1984; Babcock and Sparrowe 1989; Sparrowe and Babcock 1989). These difficulties stem from uncertainty (or disagreement) about the impacts of regulations on harvest and waterfowl abundance and from harvest-management objectives that often are vague, ambiguous, or incommensurate (Johnson et al. 1993). In the face of these ambiguities, the USFWS traditionally has taken a conservative approach to hunting regulations, thereby exacerbating the potential for conflict, particularly during periodic downturns in waterfowl abundance (Blohm 1989).

Beginning in the mid-1980s, the USFWS began searching for ways to improve the regulation of waterfowl harvests. An effort to stabilize regulations, and thus avoid much of the annual debate about appropriate regulatory responses to environmental variation, was eventually abandoned (USFWS 1988). The search for an alternative approach intensified in the 1990s when large changes in the abundance of ducks prompted renewed controversy about appropriate harvest levels. Eventually, improvements in the regulatory process were framed in terms of adaptive resource management, in which there is an explicit accounting for uncertainty as to management impacts, and for the influence of management actions on reducing that uncertainty (Williams and Johnson 1995). Since 1995, mallard hunting regulations in the U.S. have been prescribed by a formal process referred to as adaptive harvest management (Johnson et al. 1996). Efforts are now underway to extend the process to include other species of migratory game birds.

The Regulatory Process

The USFWS derives its responsibility for establishing sport-hunting regulations from the Migratory Bird Treat Act of 1918 (as amended), which implements provisions of the international treaties for migratory bird conservation. The act directs the Secretary of Agriculture to periodically adopt hunting regulations for migratory birds, "having due regard to the zones of temperature and to the distribution, abundance, economic value, breeding habits, and times and lines of migratory flight of such birds" (USFWS 1975). The responsibility for managing migratory bird harvests has since been passed to the Secretary of the Interior and the USFWS. Other legislative acts, such as the National Environmental Policy Act, the Endangered Species Act, the Administrative Procedure Act, the Freedom of Information Act, and the Regulatory Flexibility Act, provide additional responsibilities in the development of hunting regulations and help define the nature of the regulatory process (Blohm 1989).

Goals of the regulatory process are

1) to provide an opportunity to harvest a portion of certain migratory game bird populations by establishing legal hunting seasons;

2) to limit harvest of migratory game birds to levels compatible with their ability to maintain their populations;

3) to avoid the taking of endangered or threatened species so that their continued existence is not jeopardized and their conservation is enhanced;

4) to limit taking of other protected species where there is a reasonable possibility that hunting is likely to adversely affect their populations;

5) to provide equitable hunting opportunity in various parts of the country within limits imposed by abundance, migration, and distribution patterns of migratory birds; and

6) to assist, at times and in specific locations, in preventing depredations on agricultural crops by migratory game birds (USFWS 1988).

Most waterfowl hunting regulations are established annually, within a timetable that is constrained on one end by the timing of biological data collection, and on the other end by the need to give states and the public adequate opportunity for involvement before hunting seasons are established. Information on waterfowl population status, and on the outlook for annual production, is typically unavailable until early summer of each year. Some waterfowl hunting seasons open as early as mid-September, therefore, the time available for interpreting biological data, developing regulatory proposals, soliciting public comment, and for establishing and publishing hunting regulations is extremely limited. Problems or delays in the process can result in closed hunting seasons because proactive regulatory action is required to allow any harvest of migratory birds.

The annual regulatory process is documented in the Federal Register, which provides a detailed record of proposals, public comment, government responses, final regulatory guidelines, and hunting-season selections by individual states. The process includes two development schedules dedicated to early and late hunting seasons. Early seasons generally are those opening prior to 1 October and include those for migratory birds other than waterfowl (*Gruidae, Rallidae, Phalaropodidae,* and *Columbidae*) and for all migratory birds in Alaska, Puerto Rico, and the Virgin Islands. Late-season regulations pertain to most duck and goose hunting seasons, which typically begin on or after 1 October. The early-season and late-season processes occur concurrently, beginning in January and ending by late September of each year.

Early each year, the USFWS announces its intent to establish waterfowl hunting regulations and provides the schedule of public rulemaking (Figure 8-1). The director of the USFWS appoints a Migratory Bird Regulations Committee (SRC) that presides over the process and is responsible for regulatory recommendations. The SRC convenes two public meetings during summer to review biological infor-

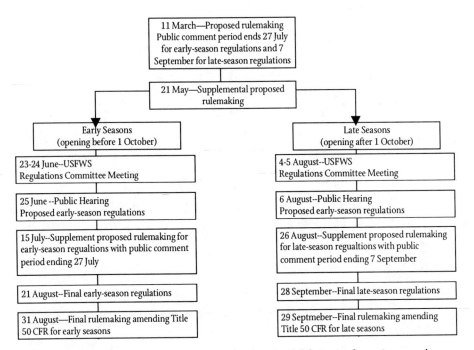

Figure 8-1 Approximate timetable used by the U.S. Fish and Wildlife Service for setting annual hunting regulations for migratory birds

mation and to consider proposals from Regulations Consultants, who represent Flyway Councils (Figure 8-2). Flyway Councils, and the state fish and wildlife agencies they represent, are essential partners in the management of migratory bird hunting. After deliberations by the SRC and Regulations Consultants, the USFWS presents hunting-season proposals at public hearings and in the *Federal Register* for comment.

Following public comment, the USFWS develops final regulatory guidelines and forwards them to the Secretary of the Interior for approval. These guidelines, referred to as framework regulations, are Flyway-specific and specify the earliest and latest dates for hunting seasons, the maximum number of days in the season, and daily bag and possession limits. States select hunting seasons within the bounds of these frameworks, usually following their own process for proposals and public comment. Final hunting regulations, including any state-imposed restrictions, are published in the *Federal Register*.

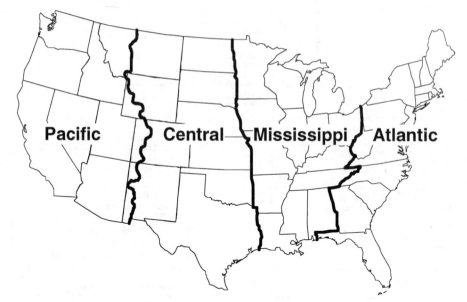

Figure 8-2 Waterfowl flyways used for administering sport-hunting regulations.

Biological Monitoring

A key component of the regulatory process consists of data collected each year on population status, habitat conditions, production, harvest levels, and other system attributes of management interest (Smith et al. 1989). This program of monitoring is essential for discerning resource status and for modifying hunting regulations in response to changes in environmental conditions. The system of waterfowl monitoring in North America is unparalleled in both scope and intensity and is made possible only by the cooperative efforts of the USFWS, the Canadian Wildlife Service, state and provincial wildlife agencies, and various research institutions. Provided here is a brief description of these monitoring programs.

Surveys conducted from fixed-wing aircraft at low altitudes are a mainstay of waterfowl management. Among the most important of these surveys are those conducted in the principal breeding range of North American ducks (Smith 1995). Each spring, duck abundance and habitat conditions are monitored in over 5 million km² of breeding habitat, using 89 thousand km of aerial transects (Figure 8-3). Ground surveys are conducted on a subset of the aerial transects to estimate the proportion of birds that are undetected from the air. The central portion of the breeding range is surveyed again in midsummer to estimate the number of duck broods and to assess the progress of the breeding season. These surveys have been operational since the 1950s and provide the most important criterion for setting annual duck-hunting regulations.

Figure 8-3 Strata and transects of the Waterfowl Breeding Population and Habitat Survey, which is conducted annually by the U.S. Fish and Wildlife Service, the Canadian Wildlife Service, and state and provincial partners

Waterfowl abundance also is determined during winter through a network of aerial surveys in the U.S. and Mexico (Smith et al. 1989). These surveys originated in the 1930s and were the basis for establishing duck-hunting regulations prior to the development of breeding-ground surveys. Winter surveys are intended to provide a census of major waterfowl concentration areas, but they lack the rigorous statistical design of breeding-ground surveys. Therefore, estimates of winter waterfowl abundance lack measures of precision and are subject to error resulting from variation in the distribution of birds relative to surveyed areas. Nonetheless, winter surveys provide useful information about large-scale waterfowl distribution and habitat conditions, and they remain the primary basis for setting most goose-hunting regulations.

Waterfowl are also monitored through a large-scale marking program where individually numbered leg bands are placed on over 350 thousand birds annually, usually just prior to the hunting season. The band inscription asks the hunter or finder of a dead bird to report the band number, date, and location to the USFWS. Banding is the principal tool used to understand migratory pathways and was the basis for establishing the four administrative flyways (Lincoln 1935). The banding program also is essential for understanding temporal and spatial variation in rates of harvest and natural mortality (Brownie et al. 1985).

The USFWS also conducts hunter surveys to determine hunting activity, harvest by species, date, location, as well as age and sex composition of the harvest (Martin and Carney 1977). This monitoring program is conducted via a mail questionnaire, which is completed by a sample of 30 to 35 thousand waterfowl hunters across the U.S.. The sampling frame is derived from purchasers of federal Migratory Bird Hunting and Conservation ("duck") Stamps at randomly selected post offices or, more recently, directly from the sale of state hunting licenses. Questionnaire results provide the basis for estimating hunting effort and total waterfowl harvest. In addition to the questionnaire, about 8,000 hunters send in wings or tails of harvested birds so that the species and demographic structure of the harvest can be determined reliably. A complete record of the waterfowl harvest in the U.S. extends back to 1962.

Predicting Regulatory Impacts

Long-term data from monitoring programs are used to estimate key population parameters, such as survival and reproductive rates, and to associate levels of harvest with various regulatory scenarios (Martin et al. 1979). These and other relevant data then are used to construct dynamic population models, which describe how waterfowl abundance varies in response to harvest and uncontrolled environmental factors (Williams and Nichols 1990). These models in turn are used to inform the regulations process by assuming that population status is directly related to harvest, and that harvest can be predicted as a function of hunting regulations (Johnson et al. 1993). By building on accumulated monitoring data, these models constantly evolve to reflect a growing understanding of waterfowl population dynamics and the impacts of harvest.

Unfortunately, the modeling of waterfowl populations and their harvest continues to be characterized by great uncertainty. In many cases, the sheer number and complexity of historic hunting regulations, combined with inadequate replication and experimental controls, has precluded reliable inference about the relationship between regulations and harvests (Nichols and Johnson 1989). Managers know even less about the impact of harvest on subsequent waterfowl population size. Particularly problematic in this regard are questions about the nature of density-dependent population regulation, which provides the theoretical basis for sustainable exploitation (Hilborn et al. 1995). It is these uncertainties about the relationships among hunting regulations, harvest, and population size that are a principal source of controversy in the regulations-setting process.

Framework for Adaptive Harvest Management

Adaptive management is management in the face of uncertainty with a focus on its reduction (Williams and Johnson 1995). In this approach, there is an explicit acknowledgment that uncertainty and risk are inherent features of natural resource management. Unlike standard approaches to risk management, however, adaptive management involves the recognition that management itself can be a useful tool for reducing uncertainty, so that long-term management performance can be improved. Adaptive management can be characterized as a problem of dual control in which managers attempt to learn about system dynamics while simultaneously pursuing traditional management objectives (Walters 1986).

In adaptive harvest management, waterfowl managers seek to maximize long-term harvest yield against a background of various sources and degrees of uncertainty (Williams et al. 1996). These sources of uncertainty are identified using the terminology of operations research and decision theory, in part to emphasize that waterfowl harvest management falls within a broad class of problems in optimal stochastic control (Nichols et al. 1995). An easily recognized source of uncertainty is uncontrolled environmental variation which produces random variation in resource status. Another source of uncertainty is partial controllability that expresses a lack of concordance between intended and actual management controls as a result of indirect actions (e.g., harvest regulations) that are imprecisely linked to specific control levels. A third source, referred to as partial observability, results from imprecision in the monitoring of harvest, population levels, and other system attributes. Finally, structural uncertainty refers to an incomplete understanding of biological processes (see also Chapters 5 and 6) and the impacts of hunting regulations. Although it is structural uncertainty that is the focus of adaptive harvest management, all sources of uncertainty influence both the ability to produce biologically acceptable harvests in the short term and to learn about system dynamics so that harvest levels can be sustained over the long term.

To account for these sources of uncertainty, adaptive harvest management was framed in terms of sequential decision-making under uncertainty, or more particularly, in terms of a stochastic control process (Puterman 1994). In this conceptual model, the manager periodically observes the state of the resource system (e.g., population size and relevant environmental features) and takes some management action (e.g., hunting regulations) (Figure 8-4). The manager receives an immediate return expressed as a function of benefits and costs that are relevant to the stated objectives of management. Based on the management action, the resource system subsequently evolves to a new state, with the transition also being influenced by uncontrolled environmental factors. The manager then observes the new system state and makes a new decision. The goal of the manager is to make a sequence of such decisions, each based on information about current system status, so as to maximize management returns over an extended time frame.

By taking advantage of the nature of decision-making and system behaviors in waterfowl harvest management, it is possible to characterize the stochastic control problem as a Markov decision process. In this class of sequential decision process, management actions, returns, and system transitions are described only in terms of current system state and action and not on states occupied or actions taken in the past. Given this simplifying constraint, computing algorithms and software are

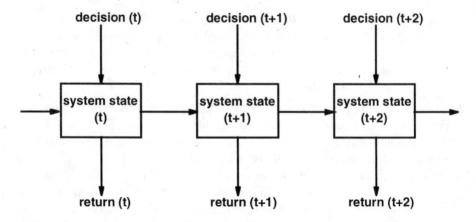

Figure 8-4 A sequential decision-making process, in which management decisions made over time (t) elicit an immediate return (benefits-costs) and then, along with uncontrolled environmental factors, drive the resource system to a new state

available for determining the optimal regulatory choice for the array of possible resource states (Williams 1996; Lubow 1995; Puterman 1994). An essential element of the optimization process is a set of state and action dependent probabilities that are associated with possible management outcomes (i.e., returns and system transitions). It is these probabilities that reflect key stochastic effects and uncertainties in system dynamics.

A major advantage of adaptive harvest management over traditional approaches is in the explicit acknowledgment of alternative hypotheses describing the effects of regulations and other environmental factors on population dynamics. These hypotheses are codified in a set of system models that are associated with a set of model-specific probabilities. These probabilities reflect the relative ability of the alternative models to describe system dynamics. Over time, some models are expected to perform better than others, and this performance is assessed by comparing the model-specific prediction of changes in population size with the actual change observed from the monitoring program. By iteratively updating model probabilities and optimizing regulatory choices, the process eventually should identify which model is most appropriate to describe the dynamics of the managed population.

Thus, the adaptive approach is a 4-step process.
   1) Each year, an optimal regulatory decision is identified based on resource status and current model probabilities.
   2) Once the decision is made, model-specific predictions for subsequent breeding-population size are determined.

3) When monitoring data become available, model probabilities are increased to the extent that observations and predictions agree and decreased to the extent that they don't agree.

4) The new set of model probabilities then are used to start another iteration of the process.

(The optimization algorithm and process for updating model probabilities is described in more detail in the Appendix to this chapter).

The key operational elements of the process include

- a set of alternative models describing population responses to harvest and uncontrolled environmental factors,
- a set of model-specific probabilities that change through time based on comparisons of predicted and observed population sizes,
- a set of alternative choices for harvest regulations, and
- an objective function by which harvest strategies can be evaluated.

These components are used to derive an optimal harvest policy that specifies the appropriate regulatory choice for various resource states and probabilities associated with the alternative models of population dynamics (Johnson et al. 1997).

The framework of adaptive harvest management has improved the regulatory process by providing a formal and coherent structure to the decision-making problem and by informing debate about appropriate levels of harvest. Unlike the traditional theory of maximum sustained yield (Beddington and May 1977), the adaptive framework accounts explicitly for the dynamic nature of ecological systems and of our understanding of those systems. The framework does have its shortcomings, however. Adaptive harvest management cannot resolve conflict over management objectives, nor can it be effective without a long-term commitment to the resource and to the pursuit of useful information about population dynamics. The adaptive harvest management process also cannot determine which management actions to consider or prescribe specific biological hypotheses. These issues demand effective institutional structures for determining how harvests should be valued by society and for ensuring productive partnerships between resource management and research.

An Example: Mallard Harvest Management

Four alternative population models capture key uncertainties (or risks) regarding the effects of harvest and environmental conditions on mallard abundance. The four models result from combinations of two discrete mortality and two discrete reproductive hypotheses (Figure 8-5). The mortality hypotheses express different views about the effects of harvest on annual survivorship. Under the additive mortality hypothesis, survival rate declines as a linear function of harvest rate.

Under the compensatory mortality hypothesis, increases in harvest rate below a threshold do not result in corresponding decreases in survivorship. The theoretical underpinning of the compensatory hypothesis is density-dependent mortality, in which mortality due to hunting is offset by declines in natural mortality. The reproductive hypotheses represent alternative views regarding the degree to which per capita reproductive rate declines with increases in mallard abundance and, thus, are also expressions of density-dependent population regulation.

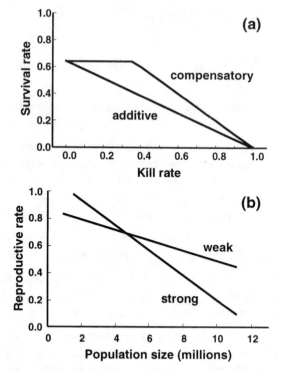

Figure 8-5 Examples of structural uncertainty: (a) hypotheses of additive and compensatory hunting mortality; and (b) hypotheses of weakly and strongly density-dependent reproductive rates

In addition to structural uncertainty, there is an explicit accounting for uncontrolled environmental variation and partial controllability of harvest rates. Stochasticity in environmental conditions is characterized by a set of probabilities assigned to various amounts of annual precipitation in southern Canada (Figure 8-6). Precipitation influences the number of available ponds which are an important determinant of mallard reproductive success. To account for

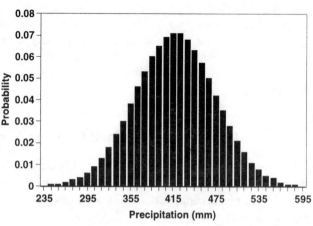

Figure 8-6 An example of environmental uncertainty: frequencies of total annual precipitation in south-central Canada over the last 50 years

partial controllability, regulations-specific probabilities are assigned to possible rates of harvest (Figure 8-7).

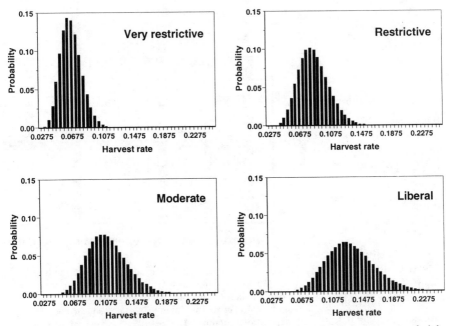

Figure 8-7 An example of partial controllability: frequency distributions for the harvest rates of adult male mallards resulting from four different sets of hunting regulations as based on past experience

Conditioned on the specification of structural uncertainty, environmental variation, and partial controllability, an optimal regulatory policy is one that is expected to maximize long-term cumulative harvest utility. Harvest utility may be defined simply as harvest yield, or as a function of harvest and other performance metrics, such as waterfowl population size. For mallards, managers seek to maximize long-term cumulative harvest, but proportionally devalue harvests whenever population size is expected to fall below the goal of the North American Waterfowl Management Plan (Johnson et al. 1996). Defining harvest utility in this way decreases the likelihood of regulatory decisions that are expected to produce population sizes below goal. Of course, harvest utility also should account for costs, but this has not been necessary in mallard harvest management because the cost of promulgating hunting regulations does not depend on the nature of the regulatory decision.

Optimal harvest regulations for mallards are highly dependent both on the status of the resource and on the probabilities associated with the alternative models of system dynamics (Figure 8-8). Regardless of model probabilities, hunting regulations become more liberal with increasing mallard and pond numbers. For a given number of mallards and ponds, optimal regulatory choices become more liberal as the probability of compensatory hunting mortality and strongly density-dependent reproduction increases.

When the adaptive harvest management (AHM) process was initiated in 1995, the four alternative models of population dynamics were considered equally, reflecting a high degree of disagreement about harvest and environmental impacts on mallard abundance. Model probabilities changed markedly

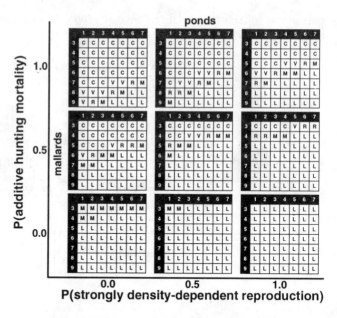

Figure 8-8 Optimal regulatory choices for mallard hunting regulations, conditioned on mallard population size (in millions), pond numbers (in millions) on the breeding grounds, and the probabilities of additive hunting mortality and strongly density-dependent reproduction (C=closed season, VR=very restrictive, R=restrictive, M=moderate, and L=liberal)

in 1996 and have remained relatively stable since (Table 8-1). On the whole, comparisons of observed and predicted population sizes provide strong evidence of additive hunting mortality and moderate evidence of strongly density-dependent reproduction. However, the set of model probabilities simply reflect the relative performance of alternative models, and conclusions regarding biological mechanisms are equivocal due to the lack of a rigorous experimental design.

Table 8-1 Year-specific probabilities associated with alternative hypotheses of mallard population dynamics

Mortality hypothesis	Reproductive hypothesis	Model probabilities			
		1995	1996	1997	1998
Additive	Strong density dependence	0.2500	0.6417	0.5668	0.6462
Additive	Weak density dependence	0.2500	0.3576	0.4235	0.3537
Compensatory	Strong density dependence	0.2500	0.0005	0.0082	0.0001
Compensatory	Weak density dependence	0.2500	0.0002	0.0015	0.0000

Relationship to the Ecological Risk Management Framework

The AHM process was conceived and implemented independently of the ERM framework described in this book. Nonetheless, the general (i.e., ERM) and specific (i.e., AHM) approaches to ERM are in conceptual agreement. Both AHM and ERM begin with a clear articulation of the management issue, including a bounding of the problem in ecological, social, and political dimensions. Both approaches acknowledge that management goals and objectives are value based, but nonetheless must be unambiguous and quantified if they are to be useful in selecting a preferred management policy or strategy. Both approaches require an a priori specification of management options or alternatives, recognizing that the set of acceptable alternatives must be limited to facilitate their assessment. Finally, both ERM and AHM depend on empirical data and its assessment to predict the ecological and social consequences of alternative management actions.

The principal difference between the ERM and AHM approaches involves the higher degree of formalism and analytical rigor in the latter. Adaptive harvest management relies heavily on the application of decision theory (Clemen 1996; Puterman 1994) in which an analytical structure provides a more systematic and objective approach to decision making. This structure is especially useful in harvest management, involving as it does sequential or dynamic decision making. Ecological management rarely involves situations in which decisions are made only once. There are many more examples where the same decision-making problem presents itself at either regular or irregular intervals (e.g., harvesting or stocking of animals, vegetation management, or water releases at a dam). The characteristic feature of a sequential decision-making process is the need to account for both current and future consequences associated with decisions made in the present. Recognizing that consequences cannot be predicted with certainty, a key difficulty in analyzing dynamic problems involves understanding how various sources of uncertainty (i.e., uncontrolled environmental variation, partial system controllability, structural uncertainty, and partial system observability) propagate over time. Fortunately, there have been recent advances in computing algorithms and software for stochastic, sequential decision-making problems, and optimal solutions for small-dimension problems now can be derived on modest desktop computers (Lubow 1995). Perhaps the most notable feature of AHM, however, is the explicit recognition that our understanding of ecological systems is also dynamic and controlled by the choice of management actions. An a priori consideration of the impacts of management choices on future levels of uncertainty distinguishes adaptive management (Walters 1986) from the more traditional tracking-and-evaluation approaches envisioned in ERM.

Summary

The Migratory Bird Treaty Act (as amended) authorizes the federal government to establish annual regulations governing the sport hunting of waterfowl within the U.S. Because of the need to collect and analyze biological data each year, the time available for developing regulatory proposals, soliciting public comment, and setting hunting seasons is extremely limited. Although the regulatory process has worked reasonably well from a biological perspective, it tends to be controversial because of uncertainties and disagreements about the impacts of regulations on harvest and waterfowl abundance. The USFWS recently developed an approach referred to as adaptive harvest management, in which managers seek to maximize long-term harvest yield against a background of various sources and degrees of uncertainty. The key feature of this approach is an explicit accounting for uncontrolled environmental variation, incomplete control over harvest levels, and key uncertainties regarding waterfowl population dynamics. Using stochastic control methodology, regulatory policies are designed to produce both short-term harvest yield, as well as the biological learning need to improve long-term management performance. This adaptive process, which has been used to regulate mallard harvests since 1995, has proved to be an effective tool for considering the relative risks of alternative management outcomes and for reducing uncertainty about regulatory impacts.

APPENDIX TO CHAPTER 8

Regulatory policies governing waterfowl harvests are identified using a recursive algorithm, in which the expected utility (or value) of harvest $V(R_t \tilde{a} X_t)$ over the time frame $* = t, t + 1, ..., T$ is conditioned on system state $\underline{X}_t$ at time t, with $\underline{R}_t$ being a policy of time-specific and state-specific regulatory decisions:

$$V(\underline{R}_t | \underline{X}_t) = \sum_i p_{i,t} \left[E\left[\sum_{\tau=t}^{T} u_{i,\tau} | \underline{X}_t \right] \right]$$

(Equation 8-1),

$$= \sum_i p_{i,t} \left[E\left[u_{i,t} | \underline{X}_t + \sum_{\tau=t+1}^{T} u_{i,\tau} | \underline{X}_t \right] \right]$$

where $u_{i,t}$ is a model-specific harvest utility and $p_{i,t}$ represents the probability that model i is the most appropriate model of system dynamics (Johnson et al. 1997). The expectation (E) is taken with respect to environmental variation and partial controllability using discrete, empirical probability distributions. An optimal regulatory policy is one that maximizes the expected cumulative harvest utility, $V(\underline{R}_t | \underline{X}_t)$.

System models that are relatively good predictors of population size gain probability mass according to Bayes Theorem:

$$p_{i,\,t+1} = \frac{p_{i,t}\, l_i(X1_t, X1_{t+1})}{\sum_i p_{i,t}\, l_i(X1_t, X1_{t+1})}$$

(Equation 8-2),

where $l_i(X1_t, X1_{t+1})$ is the probability of observed changes in population size from t to $t+1$, conditioned on model i (Hilborn and Walters 1992). This probability is calculated by assuming that observed population sizes will be distributed normally around the prediction (Williams et al. 1996; Hilborn and Walters 1992) and by deriving a simulated probability density function of predicted population size (W. Kendall, Patuxent WildlifeResearch Center, personal communication). These density functions are generated from the structure of model i and from assumed distributions for sampling variation in $\underline{X}_t$ (i.e., partial observability) and variation in harvest rates under a given regulatory decision (i.e., partial controllability).

References

Babcock KM, Sparrowe RD. 1989. Balancing expectations with reality in duck harvest management. Transactions of the North American Wildlife and Natural Resources Conference 54:594-599.

Beddington JR, May RM. 1977. Harvesting natural populations in a randomly fluctuating environment. *Science* 197:463-465.

Blohm RJ. 1989. Introduction to harvest-understanding surveys and season setting. In: Beattie KH, editor. Proceedings of the Sixth International Waterfowl Symposium. Memphis TN: Ducks Unlimited, Inc. p 118-133.

Brownie C, Anderson DR, Burnham KP, Robson DS. 1985. Statistical inference from band recovery data - A handbook. Washington DC: USFWS. Resource Publication nr 156. 305 p.

Clemen RT. 1996. Making hard decisions: An introduction to decision analysis. 2nd edition. Pacific Grove CA: Duxbury Press. 664 p.

Costanza R, Cornwell L. 1992. The 4P approach to dealing with scientific uncertainty. *Environment* 34(9):12-20, 42.

Feierabend JS. 1984. The black duck: An international resource on trial in the US. *Wildl Soc Bull* 12:128-134.

Hilborn R, Walters CJ. 1992. Quantitative fisheries stock assessment: Choice, dynamics & uncertainty. New York: Routledge, Chapman and Hall, Inc. 570 p.

Hilborn R, Walters CJ, Ludwig D. 1995. Sustainable exploitation of renewable resources. *Ann Rev Ecol Syst* 26:45-67.

Johnson FA, Moore CT, Kendall WL, Dubovsky JA, Caithamer DF, Kelley JT, Williams BK. 1997. Uncertainty and the management of mallard harvests. *J Wildl Manage* 61:203-217.

Johnson FA, Williams BK, Nichols JD, Hines JE, Kendall WL, Smith GW, Caithamer DF. 1993. Developing an adaptive management strategy for harvesting waterfowl in North America. *Trans N Am Wild Nat Resour Conf* 58:565-583.

Johnson FA, Williams BK, Schmidt PR. 1996. Adaptive decision making in waterfowl harvest and habitat management. In: Ratti JT, editor. Proceedings of the seventh international waterfowl symposium, Memphis TN: Ducks Unlimited, Inc. p 27-33.

Lincoln FC. 1935. The waterfowl flyways of North America. Washington DC: USDA Circular Nr 342. 12 p.

Lubow BC. 1995. SDP: Generalized software for solving stochastic dynamic optimization problems. *Wildl Soc Bull* 23:738-742.

Ludwig D, Hilborn R, Walters C. 1993. Uncertainty, resource exploitation, and conservation: Lessons from history. *Science* 260(4):17-36.

Martin EM, Carney SM. 1977. Population ecology of the mallard. IV. A review of duck hunting regulations, activity, and success, with special reference to the mallard. Washington DC: USFWS. Resource Publication 130. 137 p.

Martin FW, Pospahala RS, Nichols JD. 1979. Assessment and population management of North American migratory birds. In: Cairns J, Patil GP, Walters WE, editors. Environmental biomonitoring, assessment, prediction, and management - Certain case studies and related quantitative issues. Fairland MD: International Cooperative Publishing House. p 187-239.

Nichols JD, Johnson FA. 1989. Evaluation and experimentation with duck management strategies. Transactions of the North American Wildlife and Natural Resources Conference 54:566-593.

Nichols JD, Johnson FA, Williams BK. 1995. Managing North American waterfowl in the face of uncertainty. *Ann Rev Ecol Syst* 26:177-199.

Puterman ML. 1994. Markov decision processes: Discrete stochastic dynamic programming. New York NY: John Wiley and Sons, Inc. 649 p.

Smith GW. 1995. A critical review of the aerial and ground surveys of breeding waterfowl in North America. Washington DC: U.S. Department of the Interior Biological Science. Report 5. 252 p.

Smith RI, Blohm RJ, Kelly ST, Reynolds RE. 1989. Review of data bases for managing duck harvests. *Trans N Am Wild Nat Resour Conf* 54:537-544.

Sparrowe RD, Babcock KM. 1989. A turning point for duck harvest management. *Trans N Am Wild Nat Resour Conf* 54:493-495.

Teisl MF, Southwick R. 1995. The economic contributions of bird and waterfowl recreation in the U.S. during 1991. Arlington VA: Southwick Associates. 11 p.

[USFWS] U.S. Fish and Wildlife Service. 1975. Final environmental statement: Issuance of annual regulations permitting the sport hunting of migratory birds. Washington DC: U.S. Department of the Interior. 710 p.

[USFWS] U.S. Fish and Wildlife Service. 1988. Final supplemental environmental impact statement: Issuance of annual regulations permitting the sport hunting of migratory birds. Washington DC: U.S. Department of the Interior. 340 p.

Walters CJ. 1986. Adaptive management of renewable resources. New York NY: MacMillan Publishing Co. 374 p.

Williams BK. 1996. Adaptive optimization of renewable resources: Solution algorithms and a computer program. *Ecol Model* 93:101-111.

Williams BK, Johnson FA. 1995. Adaptive management and the regulation of waterfowl harvests. *Wildl Soc Bul* 23:430-436.

Williams BK, Johnson FA, Wilkins K. 1996. Uncertainty and the adaptive management of waterfowl harvests. *J Wildl Manage* 60:223-232.

Williams BK, Nichols JD. 1990. Modeling and the management of migratory birds. *Nat Resour Model* 4:273-311.

CHAPTER 9

Ecological Risk Management of a Potential Biological Stressor: The Black Carp, a Nonindigenous Species

Richard L. Orr

Introduction

The risk assessment on Black Carp was designed to test a new risk process developed by the multi-agency Aquatic Nuisance Species Task Force and has been detailed previously (Orr et al. 1999). The end point of the assessment was to determine the likeliness of the establishment of the Black Carp and, if established, the degree of negative impact that would result. Therefore, the assessment covered the ecological risk management (ERM) framework steps of issue identification, goal-setting, management options development, and data compilation and analysis, as illustrated in Chapter 2. In addition, nine specific questions ranging from the effectiveness of polyploidy in maintaining sterilization to the Black Carp's potential to feed on native mollusk populations was requested by risk managers to be answered by the assessors. The purpose of these questions was to insure that the assessment would be relevant for addressing the next steps in the risk management framework of option selection, decision implementation, and tracking and evaluation. This was done with the understanding that these final management steps were beyond the scope of the assessment proper.

The vernacular used in the Black Carp risk assessment is consistent with the terms used in evaluating invasive species (biological stressors). Most of the difficulty that goes along with linking the structure of this assessment with the terms used in the U.S. Environmental Protection Agency (USEPA) framework is because different terms are used among the different scientific disciplines. It might not be intuitively obvious that for an invasive species the probability of establishment is roughly equivalent to the USEPA's characterization of exposure and that consequences of establishment is roughly equivalent to characterization of ecological effects.

Ecological Risk-Based Decision-Making for Nonindigenous Species

Similar to addressing the risks posed by new industrial chemicals, pesticides, or waste sites, decision-making concerning nonindigenous species utilizes an interesting blend of ecological risk assessment (ERA) and ERM. Not surprisingly, ERA principles are used to evaluate biological stressors, including nonindigenous species and genetically engineered organisms. Biological stressors are unique in their ability to reproduce, adapt to new environments, and evolve over time. The concept of exposure for biological stressors includes evaluating potential entry sources and pathways (any means by which nonindigenous species are transported), as well as describing their potential for colonization and spread. In contrast to the USEPA's guidelines, which keep the ERA and ERM processes separate, but coordinated, the assessment of nonindigenous species includes risk management considerations such as socioeconomic impacts as part of their risk assessment approach.

The ERA process for nonindigenous species has been presented previously (Orr et al. 1999) and is similar in many respects to the ERA process developed by the USEPA (USEPA 1998). Plants and animals have been moved from one ecosystem to another throughout recorded history. Organisms moved outside their historic or natural geographic range are considered nonindigenous species (condition of a species being beyond its natural range or natural zone of potential dispersal). This includes all domesticated and feral species and all hybrids except for naturally occurring crosses between indigenous species (OTA 1993). Within the U.S., this includes species imported into the country, as well as those moved from one bioregion to another. In the U.S. alone, humans have intentionally or unintentionally introduced more than 4,500 foreign species that have established themselves and spread out (OTA 1993). Many introductions have been viewed as providing economic and social benefits. However, the economic and environmental consequences of some introductions have been harmful and, in a few cases, catastrophic.

Evaluations of nonindigenous species, independent of the method or process used, generally contain one or more of the following components:
- identification of one or more nonindigenous species of concern or the identification of a pathway transporting or vectoring nonindigenous species of concern,
- determination of the likelihood that these nonindigenous species could become established,
- determination of the impact if the nonindigenous species became established, and/or
- determination of the available actions to reduce the risk that the nonindigenous species will cause unacceptable damage.

The primary difference between physical/chemical ecological stressors and biological stressors is that biological stressors are capable of reproducing. Also important is the adaptability characteristic of a biological organism to control its behavior so that it can adjust to or modify the environment to fit its needs. In addition, a newly established population can, over successive generations, change (evolve) to better adapt themselves to the new environment. These basic characteristics of life add a new dimension of complexity and uncertainty that has little parallel with risk analyses on nonliving ecological stressors. The USEPA's Guidelines for Ecological Risk Assessment (USEPA 1998) attempt to incorporate these biological characteristics and provide guidelines for conducting ERA on nonindigenous species.

A number of federal agencies are involved in issues surrounding nonindigenous species. These include, but are not limited to, the U.S. Department of Agriculture, Animal and Plant Health Inspection Service; U.S. Department of the Interior, Fish and Wildlife Service; Biological Resource Division, U.S. Department of Commerce National Oceanic and Atmospheric Administration; U.S. Department of Defense; USEPA; and National Aeronautics and Space Adminsitration (NASA). A number of federal and state agencies periodically or continually conduct nonindigenous species risk assessments of varying levels of detail and sophistication for various reasons in support of their primary missions.

Federal and state governments presently share responsibilities for issues concerning the introductions of plants, animals, and their pests, parasites, and pathogens. At present, the federal effort is primarily a patchwork of laws, regulations, and policies scattered among several agencies. Most of these policies address nonindigenous species peripherally and others focus more narrowly on specific problems such as the introduction of crop pests. The need for a unifying national policy on nonindigenous species is generally acknowledged. However, the development of such a policy is impeded by historical divisions within and among government agencies and pressure from outside user groups and constituencies.

The strength of using risk assessment to evaluate nonindigenous species is that it provides a framework for taking the available information and placing it into a format that can be used and understood by policy makers for making risk management decisions. The major difficulty of using it is the high uncertainty associated with predicting the outcome of a nonindigenous species in a new environment given the lack of information on specific organisms and our current state of understanding on how an ecosystem functions. Nevertheless, the degree of uncertainty surrounding the introduction of nonindigenous organisms only increases the need for careful, unbiased risk assessments before making a decision for or against an introduction.

Even complete life-history studies of a nonindigenous species do not guarantee that managers can predict the impact that the species will have and when. The reason is that the complexity of the interaction between the organism and a new environment is so great that current predictive models do not work with enough reasonable regularity to help decision makers. Indeed, there is mounting evidence that normal linear predictive models rarely capture what occurs in a self-actualized criticality or chaotic-based ecosystem.

It is important to note that the difficulty surrounding the evaluation of an exotic biological stressor does not negate the need for management decisions to be made. It also is important to realize that because information derived from scientific methods is probabilistic and provisional, not absolute, we will never be free of uncertainty. The risk assessment, if properly designed, should allow new and innovative predictive models to be incorporated. The degree of uncertainty surrounding the introduction of nonindigenous organisms only increases the need for careful, unbiased risk assessments before making a decision for or against an introduction. It is imperative that a risk assessment honestly communicate its predictive limitations along with its strengths to policy makers.

The connection between risk assessment and risk management must be present for the risk assessment to be relevant to the needs of the risk managers. The example shows how the risk assessment (assessors) can be connected to the risk managers. The need for this type of initial bond (communication) between the assessors and the managers is recommended in the final report of the Presidential/Congressional Commission on risk assessment and risk management (1997).

References

Nico LG, Williams JD. 1996. Risk assessment on black carp (Pisces: Cyrinidae). Final draft. A report to the Risk Assessment and Management Committee of the Aquatic Nuisance Species Task Force. Washington DC. 59 p.

Orr R, McClung G, Peoples R, Williams J, Meyer M. 1999. Nonindigenous species. In: Ecological risk assessment in the federal government. Washington DC: White House National Science and Technology Council, Committee on Environment and Natural Resources. CENR-5-99-001.

[OTA] Office of Technology Assessment. 1993. Harmful non-indigenous species in the United States. Washington DC: U.S. Congress, Office of Technology Assessment. U.S. Government Printing Office. OTA-F-565.

The Presidential/Congressional Commission on Risk Assessment and Risk Management. 1997. Risk assessment and risk management in regulatory decision-making. Volume II. Final report. Washington DC: The Presidential/Congressional Commission on Risk Assessment and Risk Management. Government Printing Office. 055-000-00568-1.

Risk Assessment and Management Committee. 1996. Generic nonindigenous aquatic organisms risk analysis review process. Washington DC: Draft report to the Aquatic Nuisance Species Task Force.

[USEPA] U.S. Environmental Protection Agency. 1998. Guidelines for ecological risk assessment. Federal Register 63(93):26846-26924.

CHAPTER 10

Managing Ecological Risk Issues in a Corporate Context

Charles A. Pittinger, Ralph G. Stahl Jr., James R. Clark

The Nature of Corporate Decision-Making

The ecological risk management (ERM) framework developed at the Society of Environmental Toxicology and Chemistry (SETAC) Williamsburg VA Workshop (Chapter 2) has direct relevance to corporate management of environmental risks. Ecological risk management is similar in many ways to managing other types of business risks. Corporate risk managers are, in fact, business managers who rely upon a variety of specialized tools and technical expertise provided by others (e.g., the ecological risk assessors) to manage unavoidable risks to acceptable levels. In some cases, the corporate risk manager and the environmental manager are the same individual. A corporate risk manager may deal with environmental issues as one of many areas of responsibility, whereas an environmental manager may have a very specific responsibility for this one area alone. The issues and decisions faced by environmental managers range from product/technology development, emissions control, waste site remediation, and resource management. It is unlikely that an employee with the narrow title of ecological risk manager will be found, as those with these responsibilities are broadly trained in management including human-health and ecological risks, regulatory compliance, environmental considerations beyond risk (e.g., resource use, solid waste, life cycle efficiency, resource management), as well as the interests of the business at large. Sometimes the expertise and responsibility is vested in a group of individuals rather than a single manager.

Corporate risk managers typically face difficult choices among wide-ranging sets of options (Pittinger et al. 1998). Few decisions in a business setting offer straightforward choices. Managers balance an array of considerations in seeking safe and cost-effective solutions, including technological feasibility, social factors, and legal requirements (Chapter 1). Typically, each management option is superior in certain environmental dimensions (e.g., reducing exposure of aquatic fauna to an emission, making a material less toxic, or more rapidly biodegradable), but inferior in others (e.g., generating a hazardous waste during production, requiring more energy to implement, or being less efficacious in the product matrix). The challenge, in many cases, is not in making the commitment to do the right thing, rather, it is often

Risk Management: Ecological Risk-Based Decision-Making. Ralph G. Stahl, Jr. et al., editors.
©2001 Society of Environmental Toxicology and Chemistry (SETAC). ISBN 1-880611-26-0

deducing what is the right thing to do in a given technology development decision, as no technological innovation can be said to have zero risk (i.e., no environmental impact).

Regardless of these challenges, it is recognized that the practice of risk management in corporations is optimized when safety and regulatory criteria are integrated in decision-making together with other business demands. In product development, these demands may include such diverse considerations as consumer habits and practices, product performance and value, marketing, process feasibility and safety, efficacy and stability of ingredients in the product matrix, specifications of suppliers and contract manufacturers, labeling and handling, relationships with local and state governments and regulatory agencies, and national and international transportation and distribution networks. Likewise, in managing risks associated with waste sites and manufacturing emissions, environmental managers typically must consider emissions goals and legal requirements, demands of energy and resources, treatability of mixed wastes, and quality control considerations, in additional to balancing allotted budgets and tight time lines.

In all corporate settings, an efficient organization and decision-making process are imperative to achieve timely, high quality results. Time lines for meeting diverse needs essential to bringing a product to market or starting up a new process must be closely synchronized. To do this requires close coordination and communication among a corporation's business, risk assessment, and regulatory professionals as they evaluate all aspects of a product's life cycle (White et al. 1995).

Regardless of the activity—new product development, emissions control, waste site remediation, or resource management—the decision process benefits understanding potential ecological risks early in the decision-making process. Early development decisions tend to be less costly than late ones. However, they must be made on the basis of fewer data. Front-end loading can be very cost-effective when clear strategic thinking and available data are leveraged at the onset of a management decision. Building consensus within the organization is essential; it requires cross-functional involvement and alignment to ensure that human-health, environmental, and fiscal-management decisions are integrated vertically and horizontally within a corporation. Strong parallels exist in the regulatory arena where communicating and understanding the dimensions of acceptable risk between regulators and the regulated community are equally important.

In these ways, we view the ERM Framework as being relevant and complementary with corporate decision-making. The following describes how the framework's elements relate to and are implemented in the corporate environment.

Issue Identification

The challenge for most corporations is not so much being able to identify risk management issues, but to do so early in the process. The goal is to manage issues effectively and to do so before decisions are heavily invested. Environmental risk management issues often are characterized by the nature of the business or industry. Most industries have a suite of rather distinct and characteristic issues associated with the resources used, the production processes followed, the product's intended usage, and the wastes generated. It is our purpose here to highlight key attributes of ERM which are broadly applicable across a number of industries, recognizing that the environmental issues and management goals of different industrial sectors are quite distinct. We conclude that the need and opportunity for sound ecological risk assessment (ERA) and ERM practices are common to all.

Goal-Setting: The Role of Corporate Environmental Policy Statements

Having a clear set of environmental-quality goals and translating them into routine decision-making processes is the keystone of effective risk management. Effective ERM in a corporate setting requires three things: 1) a commitment and support of high-level managers, 2) consensus principles that are clearly articulated and utilized throughout the corporation, and 3) an infrastructure capable of making important and timely business decisions. Achieving the goals of risk management is not the sole purview of middle and senior management, as ultimately it is the employees in their day-to-day activities that actually implement the policies to achieve the goals. For this reason, a lack of employee commitment to the goals can be just as problematic as a lack of management commitment. Today, it is not uncommon for companies to publish their key environmental goals in their annual reports to stockholders, employee newsletters, and other devices to illustrate the close linkage of attaining those goals with achieving financial well-being.

Environmental quality policies provide fundamental guidance to address enigmatic social questions such as how safe is safe (enough) and how clean is clean (enough)? Some are quantitative, e.g., reducing airborne emissions of listed compounds by 50% by the year 2002. Consortia of companies in similar industries often have joint environmental management goals, such as the American Chemistry Council Responsible Care® Program (CMA 1994). Typically, these goals are implemented through a flexible process where corporate managers address legal, regulatory, economic, and social (e.g., risk perception) requirements or information relevant to their business. The policies must be sound and firm and at the same time conform with the values that society places on environmental quality.

Today, many companies regard human and ecological safety of their products and operations as a prerequisite of responsible business, not to be compromised for commercial or other reasons. Consistent with the business realities noted above, there are four fundamental needs which must be addressed by responsible environmental management (White et al. 1995):

1) human and environmental safety;
2) regulatory compliance;
3) efficient resource use and waste management; and
4) consideration of the values, needs, and concerns of society, including local, regional, and national stakeholders.

To help companies identify and implement programs, each of these needs is supported by a range of assessment tools. Protection goals for ecological systems can be broad, encompassing multiple levels of biological organization (e.g., individuals, populations, and communities). Further, these goals are applicable in a range of potentially exposed environmental media and include both structural and functional aspects of ecosystems. Efficient resource use and the satisfaction of social concerns are less tangible, but increasingly important business needs (McDaniels et al. 1997). Many corporations and trans-national organizations are wrestling with the definition and implementation of sustainable development in a business context (Shimp 1997).

Environmental goals cannot be set in isolation of business goals and market realities. Only those products and technologies which offer competitive value and improved environmental performance can survive long enough to deliver meaningful benefits to society (Hindle et al. 1993). A technology with significantly reduced environmental burdens can only deliver its environmental benefit if it is successful in the same market niche as its competition, or if it successfully eliminates that niche altogether. Any environmental benefits of a product or technology will not be realized if it renders the technology to be noncompetitive, e.g., the consumer does not want it, it wears out quickly, or it is so expensive that few can afford it. The ideal to which competitive businesses aspire is to make meaningful environmental benefits a visible and consumer-valued aspect of product innovation capable of building market share.

Similarly, the clean up of waste sites and the management of corporate resources (land in particular) also requires companies to evaluate environmental benefits and costs and to measure these against corporate environmental and financial goals. A goal for some companies is to remediate all waste sites similarly so that the properties have the potential for wide utilization after cleanup is completed. Others tailor remediation efforts to achieve particular land uses targeted in advance. For example, land intended for industrial purposes may have less stringent remediation goals than that slated for residential or parkland development. The goal of either approach is to reduce environmental exposures to contaminants to the extent necessary to protect the environment, consistent with the intended land use.

Management Options Development in a Corporate Context

The third component of the ERM framework is management options development. What options are available to the corporate risk manager to avoid ecological risks or, when unavoidable, to manage those risks to acceptable levels? Often, they are many and varied, giving the risk manager flexibility in selecting an option best suiting multiple needs. Corporate risk managers typically have available a wide and flexible array of risk management and risk mitigation options because they work directly with business units in product development, process design, and manufacturing. The engineers, chemists, scientists, and managers who devise and optimize specific technologies are uniquely capable of refining them to manage risk. Regulatory managers, by contrast, face less flexible, all or nothing decisions about using a chemical technology or are forced to make bright-line judgments based upon prescriptive legislative mandates or regulatory procedures. Their role is typically to determine whether a technology is or is not safe in its present state, rather than whether and how it can be rendered safe through effective risk management and communication.

Broad flexibility is needed to identify risk management options which satisfy changing circumstances (e.g., new sourcing strategies for ingredients, expansion of product lines, or marketing regions) that can influence the magnitude or nature of the risk. Even so, risk management options must at times be varied to meet different needs (e.g., habits and practices of product use or disposal, or waste disposal infrastructure differences around the world, varying geographical, or climatic conditions). Overly rigid or prescriptive regulations may severely inhibit innovation and curtail cost-effective risk management solutions with attendant economic consequences. For this reason, most corporations prefer regulatory criteria and processes which focus on goals rather than means, giving companies maximum flexibility to meet the goals cost-effectively and expediently.

In general, voluntary and incentive-based regulatory programs lend themselves to a achieving more lasting environmental and economic success than do command and control measures. Such measures can freeze a technology in place, making techno- logical innovation difficult to introduce. The most limiting form of imposed risk management is hazard-based (i.e., list-based) standards. These tend to rank or score chemical technologies based on inherent properties alone, treating all contexts and applications the same, and ignoring considerations of exposure. Nationalities which regulate chemical acceptability on the basis of prescribed lists of chemicals are placed at severe economic disadvantage globally. In addition, they may fail to achieve meaningful risk reduction by focusing on the most toxic chemicals and at times overlooking less recognized risks.

The corporate organization is also important in maintaining an atmosphere conducive to risk management without penalizing other corporate functions. The corporate risk manager has a wide array of risk-mitigation options available when product and process development organizations collaborate. Some vivid examples of the range of risk management options used by corporate risk managers are presented in Table 10-1. Risk-mitigation options have been characterized in detail for some industries, such as the agrochemical industry, where environmental risks are very closely scrutinized (Baker et al. 1994). Most risk management options available to corporate managers constitute a form of exposure management i.e., the production, use, or release of a chemical is controlled to ensure an adequate environmental safety margin. The same applies in the remediation of waste sites where chemical exposure can be mitigated by both natural, as well as technological, attenuation processes. In a more limited number of cases, it is possible to practice effects management where the toxicity or bioavailability of a chemical or environmental emission can be altered at the molecular level without significantly compromising efficacy, economics, compatibility, or other attributes in a product matrix.

Certain risk management options limit unacceptable applications or exposures by setting boundaries on the use of a technology, e.g., by bracketing a formula composition, limiting production output, or restricting application rates or the geographic range of the market. In some cases (e.g., test marketing), environmental managers may take advantage of staged product introductions in new markets. Beginning with smaller markets or applications, they can monitor actual environmental fate processes and exposure conditions to reduce uncertainties in the risk estimates. Other options include the establishment of criteria for sourcing raw materials (e.g., limiting impurities and unwanted byproducts), specifying production or transportation best practices (e.g., use of double-walled containers), or maintaining emergency response equipment and personnel. Still others address downstream risks of product use and disposal. Consumer education may be an important asset to emphasize preferred use and disposal practices, such as the recycling of plastic containers and used motor oil (White et al. 1995).

The final option available to the corporate risk manager, when all other options have proven unreliable or impractical, is the option of avoiding or replacing a particular technology. In this event, corporate risk managers are well advised to clearly spell out their concerns at the earliest possible stage of technology or product development to minimize cost to the company. It is common in product development to conduct screening-level risk assessments early and often in developing technologies to identify and manage unacceptable risks before significant funds are expended. In this way, companies reiterate between risk assessment and risk management in the course of the product-development process.

Data Compilation and Analysis

A variety of information and management tools are needed to ensure that diverse environmental needs and goals of a company are adequately addressed. As in other

Table 10-1 Examples of common environmental risk management approaches and practices used by corporations

Categories	Examples
Product design	-Incorporate environmental criteria (biodegradability, toxicity, etc.) in the selection of new product ingredients
	-Substitute existing ingredients with newly developed ingredients with improved environmental profiles
	-Identify novel uses of chemical technologies, and ensure risk assessments are tailored accordingly
	-Work through trade consortia to manage ecological risks common to an industry
Manufacturing	-Establish clear accountability for managing risk
	-Maintain written, widely communicated policies, SOPs, and current best approaches for manufacturing practices
	-Maintain high standards and technical specifications for raw materials to avoid or minimize undesirable impurities or by-products in raw materials (e.g., heavy metals in printing dyes; dioxin limits in paper pulp production)
	-Select or tailor manufacturing processes to avoid generation of unwanted by-products or wastes (e.g., conversion of pulp bleaching processes from elemental chlorine to hypochlorite to avoid dioxin generation)
	-Install and operated appropriate engineering controls (e.g., containers, closed piping, sumps)
	-Emergency preparedness training and procedures for spill cleanup
	-Availability of adequate and appropriate information for safe use and disposal (e.g., Material Safety Data Sheets)
	-Conducting annual safety and preparedness audits
Transportation and handling	-Evaluate safe (and potentially vulnerable) distribution or packaging practices
	-Use dedicated storage facilities and transportation equipment
	-Train personnel involved in handling procedures
	-Qualify carriers
	-Test/qualify packaging integrity under reasonable worst-case situations
	-Emergency preparedness procedures for spill cleanup
Marketing	-Product usage and disposal instructions (e.g., on packaging, through toll-free numbers)
	-Product reuse or recycling programs
	-Stage test markets and product introductions gradually to refine risk assessment, confirm assumptions, and ensure compatibility with collection, disposal and treatment infrastructures
	-Manage geographical use and release (e.g., International conventions for PBTS and Persistent Organic Pollutants)
	-Have emergency risk management procedures in place, and clearly communicated
Waste disposal and treatment	-Minimize emissions through process design
	-Segregate emissions based on type, hazard and treatability
	-Treat chemicals in situ (pump and treat, soil removal and incineration, etc.)
	-Construction of secondary and tertiary treatment plants for industrial and municipal wastes

business areas, no single tool can adequately addresses all dimensions of environmental management. Merkhofer (1999) identifies over 100 tools for the assessment, refinement, or narrowing of options. Some tools, such as risk assessment and environmental impact assessment, are based directly on environmental and health sciences and may be oriented toward regulatory compliance. Others entail more traditional business tools that provide essential information on consumer behavior, economics, and engineering technology. A number of promising new decision-making tools, which canvass several disciplines, are cost-benefit analysis, comparative-risk analysis, and cost-effectiveness analysis (USEPA 1999).

Decision-making in environmental management requires a balanced integration of the outputs from different tools, plus an interpretation based upon professional judgment and experience. Assessment tools are no substitute for professional judgment on the part of the risk manager. Prudent judgment is required to balance costs, benefits, and tradeoffs inherent in most decisions; to compensate for uncertainties due to data gaps; and to factor in variabilities in the marketplace (or, in this case, in environmental conditions) or the regulatory arena.

Sound risk management decisions require awareness of critical corporate functions beyond health, safety, and environmental as most decisions have broad ramifications. Functions which typically work in unison include product, process, and technology development; manufacturing and engineering; finance and purchasing, market research, marketing, and sales. Each function should be represented and consulted in the risk management process to ensure that foreseeable consequences and outcomes are identified and that potential adverse impacts on product or site performance are minimized. Critical time lines and pathway analyses are common tools used within industry to coordinate multiple information streams and work processes. Environmental, safety, and regulatory obligations are often represented on a project's time line with other, more conventional business functions (manufacturing, marketing, and distribution). Similar tools have also been applied in waste remediation and natural resource restoration to ensure better coordination among individuals and timely results.

For this reason, key risk managers are often situated centrally in product development, or other oversight groups within business units, where they can keep work streams flowing and eliminate potential logjams. Some environmental managers work closely with product development teams in the development, selection, and formulation of products. Others are more closely aligned to manufacturing and site operations or with remediation programs addressing past contamination at sites. A crucial role for higher levels of corporate management is to ensure that management decisions have cross-functional representation and adequate resources. They must also ensure that the decisions are consistent with corporate policies, i.e., that they can maximize business opportunities while minimizing vulnerabilities.

Option Selection: Corporate Decision-Making Criteria

Recognizing multiple connotations of "safe"

Corporate risk management decisions logically aim to reduce or avoid risk, i.e., to ensure the safety of products and operations to humans and the environment under conditions of intended use and foreseeable misuse. Yet, the meaning of the word safety varies in the eye of the beholder, whether it be the risk scientist, the regulatory official, or the lay public. There are at least three fundamental connotations: the empirical, the regulatory, and the social.

Empirical definitions of safety are synonymous with the principles of ERA espoused in this book. A process or technology might be considered as ecologically safe (i.e., carries an acceptable ecological risk) when environmental concentrations remain below adverse effect concentrations (i.e., the predicted environmental concentration [PEC] is less than the predicted no-effect concentration [PNEC]; Cowan et al. 1995). Quantitative guidelines for the use of assessment factors (otherwise called application factors or uncertainty factors) to ensure technical rigor in the ERA have been

been published by a number of authorities including the European Union (EU Commission Technical Guidance Documents) and the U.S. Environmental Protection Agency (USEPA) (Smrchek et al. 1995).

A second connotation of safe is steeped in regulatory culture. It implies that a safe technology is permissible, lawful, and compliant with relevant legal requirements. To avoid fines and penalties, the corporate risk manager needs to take into account whether and how regulatory authorities will apply such criteria in reviewing a new chemical, facility registration, or discharge permit. Beyond the satisfaction of the empirical definition, regulatory approval typically requires completion of a prescribed legal or bureaucratic process. A technology is often not deemed safe until nominal approval is granted by the regulatory body in authority.

Bureaucratic criteria are at times more challenging and unfortunately more labor intensive for corporations to satisfy than the empirical. Regulatory approval for a new technology may depend upon a variety of circumstances unrelated to risk assessment, including the completeness and timeliness of a registration procedure; the sequence through which a registration is tendered and processed; the extent to which novel approaches bypass perceived accepted practices, the progression of stakeholders weighing in on the issue, and the satisfaction of all requirements of statutory law and regulatory rulemaking.

Satisfying the regulatory prescriptions which enable a technology to be deemed safe can be an onerous challenge. Tracking and complying with changing international processes for chemical registration and approval is a formidable and expensive task for multinational corporations, particularly when regulations are inconsistent or duplicative among countries, states, provinces, or territories. Harmonization of requirements and mutual acceptance of data are needed across regions and nationalities to ensure that regulatory processes do not unnecessarily impede economic development. At the same time, harmonization should not introduce defaults to lowest common denominator to the risk-assessment process, nor should it normalize the cultural valuation dimension inherent in risk management.

A third interpretation of safe, often the most challenging for corporations to achieve, is the social or cultural connotation. Safe in this context implies a technology or a practice that seems reasonable, acceptable, or culturally appropriate to the public at large. Often the reasonableness of a decision is contingent on intangibles that are difficult to precisely assess: the perceived social value of the technology, the manner in which the risks are presented to and perceived by the public, the reputation of the company, and the surrounding circumstances of the stakeholders. A wealth of recent sociological and anthropologic research has explored factors influencing risk perception (Chapter 3). The public's concept of safety may have little to do with the practical empirical demonstration of a safety margin, or with what is legally safe from a regulatory standpoint. It may embody environmental considerations well beyond traditional endpoints, including solid waste, energy consumption, and aesthetic concerns (odor, noise). It is culturally-specific, requiring that corporations continuously strive to account for varying cultural norms and environmental conditions.

Hence, safety (and conversely, risk) should not be viewed solely in quantitative and absolute terms, but as concepts which carry different meanings in different contexts. The challenge to a corporation is to mutually satisfy all three connotations. In an era of precaution in which new technologies are often viewed with skepticism, the challenge is steep to positively balance socially perceptions of product risk with the benefits offered to consumers. Many corporations interpret the precautionary principle as being consistent with a risk–based approach to environmental management (ICCA 1995). A recent movement to broaden and encourage stakeholder participation in risk management decisions is clearly positive, yet challenges multinational companies to broaden their own understanding of how societal values and public concerns vary globally. With such a perspective, risks will be better managed and communicated to ensure that technically sound and socially acceptable safety margins are maintained. Ultimately, it is the collective values and expectations of the public which determine the acceptability of an action and indeed the commercial success of a new technology in the marketplace.

Corporate decision-making considerations to promote innovation

The complexity of environmental decision-making is why command and control environmental-management approaches sometimes are favored in government. Rigid rules based upon historical precedents and narrow compliance criteria reduce complexity, thereby simplifying a manager's life and the difficulty of a decision. Often, however, prescriptive decision-making processes overlook meaningful environmental benefits on a broader, more lasting scale. They may address only a narrow set of criteria and overlook creative and permanent solutions to larger problems. For example, there may be sound technical and environmental reasons why it may be appropriate to use an ingredient or to recycle a waste in one locality but not in another.

Likewise, scientists recognize that a chemical may be safe in one context or application, but not in another. This is the virtue and necessity of basing decisions on sound risk-assessment principles. Chemicals with relatively low toxicity but high exposure potential may represent a unique risk which could be overlooked in a hazard-based decision process. Sound risk management decisions require risk assessment and options analysis to enable a risk manager to consider circumstances and variables which may modify (either increase or decrease) risk. Sound risk management decisions require best professional judgment and a weight of evidence approach, among the many other important factors that influence the final decision.

For these reasons, a more flexible, risk-based approach is preferable to management solely according to precedence and decree. Out-of-the-box thinking leading to long-term solutions which avoid rather than prolong a problem are preferred by corporations seeking to be competitive in national and global markets. Technological innovation is fundamental to environmental improvement and is vital to finding creative solutions to current environmental problems. Effective and concerted environmental and business management encourages innovation and seeks to identify win-win solutions for all parties involved.

Post-Decision Implementation and Monitoring

Following implementation of a risk management decision, post-decision assessment is necessary in determining whether and how the decision brought about the intended change. Without such an assessment, it is impossible to determine the results of the decision and to understand the impact of future decisions. Post-decision assessment tools can be grouped generically in four classes: goals and goal systems, indicators, budgeting systems, and systematic program evaluations (Bergquist and Bergquist 1999). In a corporate context, the outcome of decisions can be traced back to a company's environmental policy to evaluate consistency with long-term goals and goal systems. The American Chemistry Council's Responsible Care (CMA 1994) initiative represents another goal system providing long-term direction and post-decision tracking. Indicators of corporate and product environmental attributes and management performance are emerging from transnational organizations such as the International Organization for Standardization's 9000 and 14000 generic management system standards (ISO 2000) and from international business consortia such as the World Business Council for Sustainable Development (WBCSD 2000).

Conclusion

This chapter sought to illustrate the complexities and commonalities encountered in undertaking ERM in a corporate setting. In many cases, the underlying framework developed at the SETAC workshop is well suited to industrial activities, whether it be for new products or technologies, waste sites, or managing corporate land. On the other hand, the diversity of concerns essential to business success in the marketplace places unique demands upon the corporate risk manager as opposed to his/her regulatory counterpart.

The practice of risk management in corporations is optimized when safety and regulatory criteria are integrated in business teams capable of integrating all essential business functions into the decision-making process. Technological innovation and flexibility in finding cost-effective ways to reduce or avoid environmental risk are highly preferred to prescriptive procedures and standards. Effective and concerted environmental and business management encourages innovation and seeks to identify win-win solutions for all parties involved. The availability of new assessment techniques decision tools, plus the growing challenge of achieving more sustainable technologies worldwide, promises to accelerate the pace at which proactive companies incorporate sound environmental risk management into routine daily business practices.

References

Baker JL, Barefoot AC, Beasley LE, Burns LA, Caulkins PP, Clark JE, Feulner RL, Giesy JP, Granel RL, Griggs RH, Jacoby HM, Laskowski DA, Maciorowski AF, Mihaich EM, Nelson Jr HP, Parrish PR, Siefert RRE, Solomon KR, van der Schalie WH, editors. 1994. Aquatic dialogue group: Pesticide risk assessment & mitigation. Pensacola FL: SETAC. 188 p.

Bergquist G, Bergquist C. 1999. Post-decision assessment In: Dale VH, English MR, editors. Tools to aid environmental decision making. New York NY: Springer-Verlag. p 285-312.

[CMA] Chemical Manufacturers Association. 1994. Responsible care overview brochure. Arlington VA: CMA.

Cowan CE, Versteeg DJ, Larson RJ, Kloepper-Sams PJ. 1995. Integrated approach for environmental assessment of new and existing substances. *Reg Toxicol Pharmacol* 21(3):3-31.

Hindle P, White PR, Minion K. 1993. Achieving real environmental improvements using value: impact assessment. *Long Range Plan* 26:36-48 p.

[ICCA] The International Council of Chemical Associations. 1995. Policy paper: Principles for risk-based decision-making. Brussels, Belgium: The International Council of Chemical Associations, c/o European Chemical Industry Council. Available on the Internet at www.icca-chem.org/issues.htm.

McDaniels TL, Axelrod LJ, Cavanagh NS, Slovic P. 1997. Perception of ecological risk to water environments. *Risk Anal* 17:341-352 p.

Merkhofer MS. 1999. Assessment, refinement and narrowing of options. In: Dale VH, English MR, editors. Tools to aid environmental decision making. New York NY: Springer-Verlag. p 231-281.

Shimp RJ. 1997. Integrating environmental management into a consumer products business. Innovation: Industry perspectives and policy implications. Proceedings of the SIGMA XI Forum; 20-21 November 1997; Washington, DC.

Pittinger CA, Cowan CE, Hindle P, Feijtel TCJ. 1998. Corporate chemical management: A risk-based approach. In: Calow P, editor. Handbook of environmental risk assessment and management. Oxford UK: Blackwell Science Ltd. p 378-400.

Smrchek J, Zeeman M, Clements R. 1995. Ecotoxicology and the assessment of chemicals at the USEPA's Office of Pollution Prevention and Toxics: Current activities and future needs. In: Pratt JR, Bowers N, Stauffer JR, editors. Making environment science. Portland OR: Ecoprint. 271 p.

[USEPA] U.S. Environmental Protection Agency. 1998. Comparative risk framework methodology and case study. SAB Review Draft. Cincinnati OH: National Center for Environmental Assessment. USEPA Office of Research and Development. Available on the Internet at www.epa.gov/nceawww1/frame.htm.

[WBCSD] World Business Council for Sustainable Development. 2000. Meeting changing expectations. Watts P, Holme R, editors. Conches-Geneva, Switzerland: World Business Council for Sustainable Development. Available on the internet at www.wbcsd.ch/printonly.htm#top.

White PR, De Smet B, Owens JW, Hindle P. 1995. Environmental management in an international consumer goods company. *Resourc Conserv Recyc* 14:171-184.

CHAPTER 11

Managing Ecological Risks and Restoration of a Wetland Habitat at a Superfund Site: A Retrospective Assessment of the Proposed Ecological Risk Management Framework

Ralph G. Stahl Jr., Maryann J. Nicholson, P. Brandt Butler, John E. Auger, Edward J. Lutz, John E. Vidumsky, Pamela Meitner

Introduction

The ability to make reasoned, timely decisions on remediating contaminated sites is a highly desirable, but often lacking, element of ecological risk management (ERM). Making a decision on which action to take seems to become mired in protracted and overly detailed discussions particularly when ecological risk assessment and ERM are involved. In some cases, the slow pace may result from the lack of clarity by the decision-maker in articulating exactly what information is needed compared to the information that the scientist, risk assessor, or engineer is willing/able to offer. Quantitative analysis of potential risk management options is often not conducted or, if done, it is laced with high uncertainty. The latter was a key point raised in the Society of Environmental Toxicology and Chemistry (SETAC) sponsored Workshop on Framework for Ecological Risk Management 23-25 June 1997 in Willamsburg, VA (Chapter 15; Pittinger et al. 1998) and was no less important in the case of this DuPont Superfund site (DSS).

The DSS is situated along a tidally-influenced tributary of a major northeastern river and had been actively manufacturing a diverse array of products for over 40 years prior to its closure. In the 1930s and 1940s, off-spec product and some wastes were landfilled in a nearby wetlands. Some of the wastes contained barium and other metals that entered the surrounding wetlands and river systems and became the constituents of concern (COC) that led the U.S. Environmental Protection Agency (USEPA) to list the site on the National Priorities List.

In 1993, the USEPA issued DuPont a Record of Decision (ROD) that provided the rationale for addressing the risks to human health and the environment and the remedy that would be required. The primary impetus for the ROD was remediating potential risks to ecological receptors located in the wetland and river habitats near the site. Risks to human health were much less of an issue and are not discussed in this case study. Although several components of the ROD required remedy, the one underlying this case study was excavating contaminated sediments in a wetland area, followed by restoration of the excavated area to grade using clean fill material.

Since 1991, DuPont has utilized a corporate-level approach to remediating waste sites in North America and Europe. This centralization of the effort spawns economy of scale and allows key learnings and experience to be leveraged across a wide variety of situations. The DSS represented the first site where risks to ecological receptors, and the need for ERM, was encountered. Because of the complexity of the issues associated with this site, and the need for wide expertise to implement the remedy, an internal core team of DuPont engineers, risk assessors, lawyers, environmental scientists, and other experts from outside the company was established. Members of the internal team met regularly with USEPA and state regulators to discuss how best to implement the remedy. When restoration of the habitats and final discussion of the ERM decision arose, individuals from the National Oceanic and Atmospheric Administration and the U.S. Fish and Wildlife Service also participated. Although it was not available to the team during this process, the proposed ERM framework has many steps in common with those used by the DuPont team. For this reason, the efforts at the DSS are discussed retrospectively in the context of the proposed ERM framework shown in Chapter 2 to provide a tangible example of the potential utility of its seven steps.

A Retrospective Application of the Proposed Ecological Risk Management Framework Issue Identification

A phased sampling approach was valuable in fully delineating the extent of COCs in the wetland habitat. This allowed the team to define more clearly the potential excavation footprint and achieve a more comprehensive understanding of what the ROD-required remedy might or might not accomplish in this area. One of the first issues became how much more of the COC-contaminated sediment could be removed without incurring substantial additional cost, perhaps providing a greater level of risk reduction than originally required. The concept was that removing additional sediment, beyond that required by the ROD, would reduce the scope of any maintenance and monitoring activities as the likelihood of COCs remaining in the environment would be minimized. This was joined by a second issue of determining whether there were restoration options that could be implemented during remedial efforts that would, without incurring substantial additional cost, increase the ecological services provided by the restored wetlands above and beyond the requirements of the ROD. To address these two issues, the internal team was faced with, yet, a third issue. This was the need to develop quantitative financial and ecological measures that would allow a critical, quantitative evaluation of the various risk management and restoration options.

Goal-setting

After much discussion internally, and interaction with the USEPA and others, the internal team developed a series of broad goals that were reviewed at each meeting and shared with stakeholders and upper management. These included meeting the legal requirements of the ROD, gaining agency and public acceptance of remedial and restoration decisions, conducting all work safely and cost-effectively while being protective of human health and the environment, and, finally, basing decisions on sound science and good information. Although it was not listed in the above set of goals, another more specific goal was to provide an increased level of ecological services in the restored habitats by implementing a remedy that would exceed the requirements of the ROD, at little or no additional cost.

The implementation of the remedy would proceed in three stages: a small wetland area followed by another, larger wetland area, and ending with remedial activities in the nearby river. The first test of the decision-making process began with the small (about 2 acres) wetland habitat. To the internal team, success in this small area would indicate that the approach could be applied to the larger wetland area (nearly 10 acres in size) and perhaps to the river itself. Likewise, a successful approach in these areas might prove applicable and beneficial at other DuPont sites.

Management options development

Sediments of the tidally influenced wetland were contaminated with the COCs and the habitat heavily dominated by the exotic and invasive *Phragmites australis* (common reed). All risk management options developed by the internal team and support personnel encompassed excavation of the contaminated sediments, coupled with varying degrees of restoration and habitat enhancement. Not only were there varying degrees of restoration afforded among the options, there were varying degrees of excavation that ranged from the required 12 inches (the ROD requirement, also called the base case) to a removal of sediment down to the historic marsh deposit (2 to 3 feet). As each of the options was conceptualized, it was added to matrix tables that were used to illustrate graphically the range of potential actions (Tables 11-1, 11-2, 11-3). The matrix tables became a valuable tool to highlight data gaps and communicate with upper management.

Table 11-1   Functional capacity index

Function	Current	ROD	Optimal
Shore / bank	0.31	0.31	0.62
Sediment stabilization	0.75	0.75	0.8
Water quality	0.45	0.68	0.8
Wildlife habitat	0.19	0.50	0.85
Fish habitat	0.56	0.59	0.75
Uniqueness	0.90	0.90	0.90

Data compilation and analysis

After the conceptual work was completed and the risk management options applied to the tabular matrix, it was necessary to develop the quantitative financial and ecological measures to support each option. This information could be added to each option so that decision-makers could judge the merits of each against the other. Engineers on the team were tasked with developing costs for the options based on standard engineering cost estimation methods. This group was also required to view the remediation

Table 11-2 Functional capacity units

Function	Area (ac)	Current	ROD
Shore / bank	0.18	0.056	0.056
Sediment stabilization	2.5	1.9	1.9
Water quality	2.5	1.13	1.7
Wildlife habitat	2.5	0.5	1.25
Fish habitat	2.5	1.4	1.5
Uniqueness	N/A	N/A	N/A

Table 11-3 Ecological risk management and restoration options matrix

Action		Rationale
1.	Excavate / replace sediments	ROD – base case
2.	Create pools and channels	Improve fish habitat quality
3.	Create hummocks and islands	Improve wildlife habitat quality
4.	Control erosion	Maintain restoration areas
5.	Control invasive Phragmites	Improve plant community
6.	Combine alternatives	Achieve best mix for lowest cost

in a holistic manner because restoration was being coupled with removal activities. For example, a road was to be constructed to allow access for equipment into the wetlands and as a way to remove sediments by truck. After remediation, the road could be altered to provide hummocks, islands, etc. within the restored wetland rather than having to remove it once the remediation was completed.

Similarly, the scientists and risk assessors on the team were tasked with developing quantitative measures of ecological services that would not require extensive, lengthy data collection. The plan was to use existing information where possible, yet gather additional data when uncertainties were deemed too great. An underlying assumption by the risk assessors and scientists was that all the risk management options under consideration would provide equal or better protection to the environment than the actions required by the ROD. Because the first area addressed was the small wetland area, it was necessary to evaluate wetland restoration protocols for their ability to provide the quantitative measures of ecological services that would be needed. After reviewing three protocols, The Evaluation for Planned Wetlands (EPW) (Bartoldus et al. 1994) was selected because it provided a relatively simple, yet quantitative, approach to valuing the ecological services (functional attributes) of various types of wetlands. The methodology allowed the functional evaluation of the existing wetland, as well as the planned or restored wetland, and was sensitive enough to detect differences between the existing and restored areas. Each of the six categories of wetland functions contained in the EPW approach could be assessed and scored numerically without the need for extensive data collection. The EPW approach first results in an estimate which describes the wetland's relative capacity to perform each function (functional capacity index)

(Table 11-1). The second step is to factor in the size of the area where the service or function is provided, resulting in the derivation of a wetland-service-unit (functional capacity unit) (Table 11-2).

Once the wetland-service-unit was derived for the various areas, enhancements and changes to those areas using various restoration actions could be fed into the model to derive the incremental improvement in ecological services. Coupled with the engineering cost estimates attached to each of these enhancements, the internal team now had a simplified method to quantitatively illustrate the cost of an enhancement with the gain in ecological service (Table 11-4). All enhancements were judged against the base case, or ROD-required remedy and restoration. The model could be run with differing combinations of excavation and restoration, providing a useful tool for rapid evaluation of small and large changes alike. A more detailed description of this methodology for publication purposes is currently under development.

Table 11-4 Ecological and financial cost/benefit analysis of risk management and restoration options

Action	Net increment improvement over current condition	Cost/acre	Cost/increment improvement
ROD (base case)	5	$ 53,000	$ 10,600
Construct pools/channels	3	$ 68,000	$ 22,600
Construct hummocks/islands	1	$ 66,000	$ 66,000
Control erosion	5	$ 6,200	$ 1,240
Control phragmites	0.4	$ 1,000	$ 2,500

Option selection

Selecting the appropriate risk management and restoration option required a number of candid, often spirited meetings among the core team where all the influences on the ERM decision (see Chapter 1) were encountered. These discussions were no doubt a microcosm of the many that ensue when addressing ERM at waste sites or elsewhere for that matter. From the numerous possible risk management options that employed a combination of excavation and restoration actions, the one selected ultimately was not the least expensive, nor the most easily implemented. What it did provide was enhanced ecological services and from a longer term perspective was more protective than the actions specified in the ROD. Another key selection criteria that helped to eliminate some potential options was evaluating the future needs of maintenance and monitoring. Remediating and restoring a wetland is not a construct and leave situation. Maintenance and monitoring are needed to ensure that vegetation used in the restoration is taking root and growing and that erosion is repaired so that the restored grade and vegetation are not damaged. Implementing an option that reduced the scope of the maintenance and monitoring activities would translate into future savings. Having a sound design and a regularly monitored implementation program can help to rectify small problems before they become large and costly.

Decision implementation

Once the internal team was sure the costs were appropriate and the preferred risk management and restoration option feasible, it was presented to the USEPA and others involved with the DSS. In actuality, informal discussions with regulators were held throughout the option development and selection process. When the preferred option was presented, there was general agreement that it was an improvement over the original remedy and restoration required in the ROD. The preferred option resulted in a near doubling of the sediment removed, as well as a significant improvement in ecological services as quantified by the EPW approach. Although some changes to the restoration were made based on discussions with scientists and risk assessors from the regulatory agencies, in general the preferred option was accepted with little modification.

Tracking and monitoring

The preferred option for remediating and restoring the wetland area was implemented in mid-1997. A maintenance and monitoring plan was developed and implemented for the wetland based primarily on ensuring revegetation of the wetlands and stopping erosion before it damaged the habitat. Over the 4 years since implementation, the wetland area has revegetated appropriately and no erosion has been observed. As many wetland restorationists have found, major impediments to wetland restoration are resident and migratory waterfowl that view newly planted wetland plants as a prime food source (Kusler and Kentula 1990). Waterfowl were problematic in the initial stages of revegetation of this wetland, but their impact (feeding on newly planted vegetation) was reduced by employing simple exclusion techniques (goose fence).

Summary

This case study was provided to illustrate retrospectively how the seven steps in the proposed ERM framework could be applied to selecting an appropriate risk management option for remediating and restoring a wetland area at a Superfund site. Steps taken by the internal DuPont team were similar to those provided in the proposed ERM framework, illustrating the potential utility of the framework. The wetland area at the DSS was the subject of a public celebration by the regulatory agencies and various DuPont personnel in March 1998. It was hailed by speakers from the regulatory and regulated groups as a model for addressing Superfund sites. As this book goes to press in mid 2001, visual observations confirm that the wetland is thriving and the habitat widely used by fish, waterfowl, and numerous other wildlife within the watershed.

Acknowledgments -The authors wish to thank the numerous individuals in the regulatory community who were key to the success of this project. Drs. John Parker and C. Michael Swindoll were important contributors to the wetland field assessments and conceptual discussions. Mr. Matt Brill provided geological expertise and advice for the remedy and restoration. Mr. Keith Rudy provided important engineering cost estimates and design options. Mr. William Pew was instrumental in the final engineering design and implementation of the wetland remediation and restoration. We thank Ms. Patricia Quigley, Mr. Edward Garbisch and Mr. Mark Krause who helped design the wetland restoration and provided useful advice on how to ensure it was successful. Finally, we thank the other scientists and engineers in DuPont who worked with the internal team on designing and implementing this remedy.

References

Bartoldus CC, Garbisch EW, Kraus ML. 1994. Evaluation for planned wetlands. St. Michaels MD: Environmental Concern Inc.

Kusler JA, Kentula ME. 1990. Wetland creation and restoration: The status of the science. Washington DC: Island Press.

Pittinger CA, Bachman R, Barton AL, Clark JR, deFur PL, Ells SJ, Slimak MW, Stahl RG, Wentsel RS. 1998. A multi-stakeholder framework for ecological risk management: Summary from a SETAC Technical Workshop. Pensacola FL: SETAC. 24 p.

CHAPTER 12

Ecological Risk Management and Restoration of Estuarine Habitat Following a Chemical Release: The Calcasieu Estuary Restoration Project

Andrew L. Strong

Introduction

This Chapter discusses the integrated management of ecological risks and natural resource injuries through a common denominator—habitat restoration—and in so doing, it follows a number of tenets detailed in the ecological risk management (ERM) framework (Chapter 2). As risk is theoretical, injury is actual. Or, in other words, risk means risk of injury. While often stemming from differing statutory authorities, the approaches and types of data collected in ecological risk assessments (ERA) and natural resource damage assessments (NRDA) are quite similar in many respects. Both typically focus on the same ecological receptors and the same habitat and look prospectively to determine risk and injury. The most glaring difference, however, is that an NRDA might address past, present, and future injuries, whereas an ERA is usually only prospective in nature. Additionally, an NRDA might address injured resources used by humans such as groundwater, parks, and fishing areas. Nonetheless, because of the similarities, when addressing ecological risk issues, it can be desirable to also address natural resource injuries in order to cost effectively manage both liabilities. The following case example supports this approach. However, all sites are different and the strategy for managing ecological risks and natural resource injuries should be evaluated based upon site-specific technical, legal, and political factors.

Issue Identification

In March 1994, an episodic release of approximately 1.6 million pounds of ethylene dichloride (EDC) occurred in a maintenance ditch in Lake Charles, Louisiana. A portion of the amount released flowed into the northern portion of Clooney Island Loop (the Loop), an arm off of the Calcasieu River and part of the Calcasieu Estuary. As EDC is heavier than water, free-phase product flowed into the Loop and accumulated in low spots in the immediate vicinity of a barge/tanker loading dock. Emergency response activities began after discovery of the release and concluded approximately 9 months later.

Goal-setting

The goal of the response activities was to remove the free-phase EDC and affected sediment to levels less than or equal to 1,000 mg/Kg EDC. This action was driven by the need to reduce or mitigate continued exposure of ecological receptors to the EDC in sediments and the water column.

Management Options Development

To achieve the above goal, substantial resources were committed by employing large dredging and excavation equipment and creating an elaborate materials-management process. The latter included water and sediment management equipment such as dewatering tanks, thermal desorption units, carbon filtration, and water purification units. All told, the estimated cost of the emergency response activities for the 9 month period was over $20 million.

This activity did not follow exactly the ERM framework due to the requirement for immediate action. In this case, steps 4 and 5 of the framework were skipped and the team simply went to step 6, decision implementation.

Additional Management Options Development and Data Compilation

Following the emergency response activities, an evaluation was performed to determine whether additional remedial action of the affected sediment in the Loop was merited. That evaluation involved a screening-level assessment of potential human-health and ecological risks, predictive transport modeling of EDC in the sediment and water column, and the potential ecological benefits of conducting additional sediment remediation. The risk assessment and modeling work indicated that the human-health and ecological risks were temporal and would dissipate over the next 18 months, thereafter meeting Louisiana's desired risk-reduction goal. Moreover, because of the highly anoxic and hypoxic conditions along the water/sediment interface in the Loop, the benefits of additional remedial action were nominal. That notwithstanding, elevated levels of EDC in the sediment remained, albeit temporarily, and arguably some level of ecological risks to benthos, fish, and wading and fish-eating birds existed. Moreover, social and political pressures were mounting for additional action to be taken, regardless of technical prudence.

From an NRDA perspective, based upon this technical analysis, it could be argued that the potential for natural resource injuries to the same ecological receptors and habitat in the Loop was small. However, proving the amount of injury (also referred to as the debit) would require additional costly studies and data review and possibly lead to lengthy delays in the final resolution of the matter. Thus, with the prospect of having to conduct additional studies, combined with the uncertainties in the ERA, it was proposed to the remedial and trustee agencies that both the ecological risks and natural-resource injuries be managed together through the use of a restoration-based approach. That approach involved the use of habitat restoration to compensate for all future risks and past and future natural resource injuries affected by the release of EDC and corresponding remedial activities. The goal of the habitat restoration was focused on restoration activities supporting the same or similar ecological receptors and natural resources affected by the EDC release.

In order to determine the minimum amount of restoration necessary to compensate for the ecological risks and natural resource injuries, an economic algorithm referred to as the Habitat Equivalency Analysis (HEA) model was used. The HEA model was originally developed by the Department of the Interior and the National Oceanic and Atmospheric Administration and is used to assess the scope of the injury (debit) and the amount of compensatory restoration (credit) through a common denominator —discounted service-acre-years (dSAYs)[1]. Simply put, the debit is the lost ecological services from a habitat resulting from the release of hazardous substances or oil to that habitat. A habitat, such as a wetland area, provides many services for wildlife and aquatic life such as foraging area, shelter, reproduction, nesting, as well as water quality filtration from runoff. A decrease in these services attributable to a release of hazardous substances or oil to the habitat can be computed using the HEA model. The key debit inputs to the model are 1) the area affected by the release of hazardous substances or oil (e.g., spatial extent of sediment exceeding risk-based values or oiled vegetation); 2) the percent loss in ecological services from a baseline (the services provided by the habitat, but for the release of hazardous substances or oil); and 3) the duration of the service loss (past and/or future). The credit is the amount of restoration required necessary to offset the debit.

Since the remediation activities focused on the affected site, an off-site restoration project, within the same watershed, was selected. An early attempt was made to select a site with identical habitat to that which was affected in the Loop (predominantly dredged river-bottom sediment and open water habitat), however, that effort was quickly deemed inappropriate because it is technologically infeasible to restore and/or rehabilitate an identical off-site habitat because of the naturally occurring conditions (anoxic/hypoxic) and commercial activities present in the Loop; and the injured habitat is low in biological diversity and density, the restoration of an identical habitat is undesirable and would likely not yield significant ecological benefits. Thus, a similar, but not identical, habitat (e.g., soft-bottom sediment in a wetland area) was used as the biological metric. It was generally agreed that this

---

1 dSAY is a unit of measurement that is used to assign a numerical value to the amount of ecological services provided by one acre of habitat in one year, as discounted over time.

type of habitat was of greater value and quality than the Loop's habitat and restoration of soft-bottom sediment wetlands in the Gulf Coast has a proven track record.

To properly scale the proposed restoration activities, the HEA model was used. Based on data and information collected during the emergency response activities, the EDC release potentially affected 9 acres in Clooney Loop and 4 acres in the drainage ditch, totaling 13 acres. While the modeling showed the EDC concentrations dissipating after 18 months, the temporal estimate used in the HEA model was 5 years. It was also assumed that the lost services for this time period was 100 percent (this assumes that injury and risk within the 13 acres is certain for the 5 year period).

Using these input assumptions, the HEA model computed the debit to be 20.7 dSAYs. Thus, in order to demonstrate that the proposed restoration activities adequately addresses the ecological risks and compensates for the natural resource injuries, the credit must simply be greater than or equal to the debit, i.e., the restoration project must generate at least 20.7 dSAYs.

However, even if the restoration project satisfied this credit number, there were still potential uncertainties in the ERA and fate and transport modeling, agency concerns regarding the use of differing habitats, and political and community related concerns regarding the affected site (see also Chapter 3). To address these concerns, there were two options: 1) conduct more studies to further support the ERA and the rationale for leaving the EDC in the sediment, or 2) increase the size of the restoration project so that there was no question that it adequately compensated for the risk/injury and any uncertainties attached thereto. The latter option was selected. As restoration costs are significantly less than remediation costs, one can afford to be generous. In this case, it was estimated that further remedial action at the Loop, if required, could surpass the costs incurred during the emergency response activities (more than $20 million). In addition, because of the political and public concerns regarding the release, while not necessarily technically-based, a restoration project of significant size and uniqueness was important to the ultimate success of the resolution.

Therefore, rather than look for a project that simply generated enough credits to slightly exceed the debit, a much larger project that had great ecological and public appeal was identified. The HEA model was then used to demonstrate that the credits that would be generated by the project would far surpass the debit (the credit was estimated to be 241.2 dSAYs, over 10 times the debit). The project was quite large because it involved restoration of over 200 acres of a intertidal marsh and bottomland hardwood ecosystem that had been completely denuded by the current owner and used as pastureland for cattle. The proposed restoration project would provide sanctuary to many indigenous animals and fish species. In addition, more than 50,000 one-year-old tree saplings would be planted, including species such as the bald cypress, overcup oak, nuttal oak, green ash, bitter pecan, common persimmon, willow oak, and water oak.

Option Selection and Decision Implementation

Based upon the HEA analyses and the high value of the proposed restoration project, the remedial and trustee agencies and the public agreed that the completion of the proposed restoration project would more than adequately address potential ecological risks and natural resource injuries at the affected site. It was, therefore, agreed that the restoration project would be conducted in lieu of additional dredging at the affected site so long as conditions at the affected site continued to improve (as predicted in the fate and transport model). That agreement was reached in 1996, and a consent decree formalizing the agreement was entered by the federal district court in 1997. The restoration project also commenced in 1997, and many of the tree saplings are now over 6 feet tall.

The restoration project has produced positive press coverage and has resulted and will continue to result in multiple wins for the public, regulators, responsible party, and, most importantly, the environment. Moreover, a 2-year thesis study by a graduate student from Louisiana State University is underway to document the success of the project and to aid in future restoration projects in the Gulf Coast.

Tracking and Evaluation

As to conditions in the Loop, several years of water quality monitoring and sampling since 1996 confirmed that the fate and transport model was overly conservative. The chemical sampling results were less than a tenth of the allowable concentration originally established by the agencies in the agreement.

Summary

There were many elements of the ERM framework applied in this case involving the release of a chemical into a estuarine environment. However, it should be noted that due to the need to implement a remedial activity quickly, some steps in the ERM framework were skipped. Looking to the future, it is clear that the ERM framework is applicable to numerous situations where time is sufficient to allow careful development of options and their selection. In some cases though, the need to act will be acute and may not allow scientists and decision-makers the time to utilize all steps in the ERM framework.

CHAPTER 13

The Nature Conservancy's Five-S Decision-Making Framework for Site-Based Conservation and Relationship to the Proposed Ecological Risk Management Framework

Jeffrey V. Baumgartner

Introduction

Conservation of biodiversity is a challenging undertaking. Biodiversity can be defined at multiple scales of organization, from genes to landscapes (Noss 1990). As the focus of conservation shifts from single species to larger scale approaches (Poiani et al. 1999; Franklin 1993), our ability to understand and address the complexities of dynamic ecological systems and landscapes is challenged. We also must characterize and address the complex interactions between human communities and ecological systems. Finally, with the acute nature of the biodiversity crisis and the relatively limited resources committed to biodiversity conservation, we must make timely strategic decisions about what biodiversity to conserve, where to conserve it, and how to conserve it most efficiently and effectively. This paper presents a conservation decision-making framework developed and used by The Nature Conservancy that balances the need for scientific rigor with the need to take quick and effective action. Conceptually, this framework parallels the ecological risk management (ERM) framework presented in Chapter 2.

The Nature Conservancy is a nonprofit, conservation organization that has been in the business of conserving biodiversity for nearly 50 years. The Conservancy's conservation goal is the long-term survival of all viable native species and ecological communities through the design and conservation of portfolios of sites within ecoregions. The locus of our conservation work is sites—the places on the landscape where we and our partners are working to protect specific elements of biodiversity (i.e., species, ecological communities, and ecological systems). The Conservancy's ecoregional approach to conservation (The Nature Conservancy 2000a) has four major components: 1) ecoregional conservation planning, or site selection; 2) site conservation planning, or site design; 3) strategy implementation; and 4) measures of conservation success. The second component, site conservation planning, is the focus of this paper. Below is the basic framework for site conservation planning. Alternative and more detailed presentations of this approach to site conservation

---

Risk Management: Ecological Risk-Based Decision-Making. Ralph G. Stahl, Jr. et al., editors.
©2001 Society of Environmental Toxicology and Chemistry (SETAC). ISBN 1-880611-26-0

can be found in The Nature Conservancy (2000b), Poiani et al. (1998), and Weeks (1997).

The Five-S Framework for Site Conservation

Every conservation site has one or more prima facie reasons it has been selected for conservation—i.e., occurrences of important species, ecological communities, or more broadly defined assemblages of species and communities. These species, communities, and assemblages are referred to as conservation targets. Site conservation planning is a decision-making and problem-solving process for effectively and efficiently conserving the conservation targets at a site. Assessment of the ecological characteristics of the targets (i.e., the ecological patterns and processes that sustain the targets) and the human context (e.g., cultural, economic, legal, and political issues) leads to the development of conservation strategies designed to abate critical threats, maintain the ecological characteristics, and meet human community needs. The planning framework has five major components, referred to as the five S's, including

- systems—the conservation targets at a site, and the natural processes that maintain them;
- stresses—the types of destruction, degradation, and impairment afflicting the systems;
- sources—the agents generating the stresses;
- strategies—the types of conservation activities deployed to abate sources of stress (threat abatement) and maintain, enhance, or restore the systems (restoration); and
- success—measures of biodiversity health and threat abatement at a site.

The logic underlying the five-S framework is simple. The goal of site conservation is to maintain healthy, viable occurrences of the conservation targets. By definition, healthy, viable occurrences are not significantly stressed. Logically, abating the sources of stress should alleviate the stresses and, consequently, enhance the viability of the systems. In those cases where it is not feasible to abate a source, or when the stress persists after the source is removed, direct restoration or enhancement of the conservation targets through active management may be warranted. Thus, conservation strategies are developed and implemented to abate the critical sources of stress and directly restore, enhance, or maintain the systems. The measures of conservation success then assess the effectiveness of our strategies.

Collectively, the first three S's (systems, stresses, and sources) correspond to the first two components of the ERM framework (issue identification, and goal-setting). The strategies component of the five-S framework is analogous to the management options development and option selection components of ERM, as is success to tracking and evaluation. Although information compilation is not an explicit component of the five-S framework, it is implicit. Application of the framework is dependent on sound information about ecological and human systems at the site.

Systems

Site conservation begins with identifying and characterizing the priority systems (conservation targets) at the site, including the natural processes that maintain them. The Nature Conservancy generally considers 3 types of conservation targets: species, ecological communities, and ecological systems. Species targets include imperiled and endangered native species; species of special concern due to vulnerability, declining trends, disjunction, or endemic status; major grouping of species that share common natural processes or have similar conservation requirements (e.g., freshwater mussels, forest-interior birds); and globally significant examples of species aggregation. Ecological communities are groupings of cooccurring species, as defined at the "association" level of the Conservancy's U.S. Nations Vegetation Classification (Grossman et al. 1998) and the "alliance" level of the aquatic community classification (The Nature Conservancy 1997). Likewise, ecological communities may be aggregated into ecological systems, dynamic assemblages that

1) can occur together on the landscape;
2) are tied together by similar ecological processes, underlying environmental features (e.g., soils, geology) or environmental gradients (e.g., elevation); and
3) can form a robust, cohesive, and distinguishable unit on the ground.

Ecological systems can be terrestrial, aquatic, marine, or combination of them.

For The Nature Conservancy, conservation targets typically are identified through a site selection process (e.g., ecoregional planning). For others involved in site-based conservation (e.g., federal land management agencies), it may be a policy mandate (e.g., Endangered Species Act). Identification of conservation targets is the basis for all subsequent steps in site planning, including identifying threats, developing strategies, measuring success, and delineating the site boundary. A different set of targets is likely to result in different threats, strategies, measures of success, and site boundaries. Although identifying and characterizing systems may be perceived as an easy task, in practice, it is often the most difficult.

It is important to have a manageable number of conservation targets as the focus for site planning. The conservation targets at a site, as identified through the site selection process or otherwise, may be too numerous to practically address individually. Also, it is often more efficient to focus on higher functional levels of biodiversity organization, one or a few ecological systems rather than many individual species and ecological communities (Franklin 1993; Noss 1996; Poiani et al. 1999). Thus, it is important to screen the conservation targets for a small subset (e.g., no more than eight) of priority targets that will be the focus for site planning.

To ensure that the priority conservation targets sufficiently represent the broader array of conservation targets at the site, screening for priority targets is done at multiple scales of biodiversity organization, beginning at higher levels and working down.

- Screen for ecological communities and systems at the coarsest scale (matrix ecosystems; Poiani et al. 1999);
- Consolidate individual species and ecological communities into major groupings and ecological systems, respectively, and divide large landscapes into major ecological systems;
- Identify individual species or ecological communities that have special requirement; and
- Identify individual species or communities that integrate across ecological systems.

Once the priority conservation targets have been identified, it is necessary to characterize the ecological context in which they exist at the site. Maintaining the viability of the priority targets at the site will depend upon maintaining the ecological processes and interactions that allowed them to establish and thrive in the past. Three factors—size, condition, and landscape context—should be considered in characterizing the ecological context of the targets.

Size is a measure of the area or abundance of the conservation target's occurrence. For ecological systems and communities, size is simply a measure of the occurrence's patch size or geographic coverage. For animal and plant species, size takes into account the area of occupancy and number of individuals. Minimum dynamic area, or the area needed to ensure survival or reestablishment of a target after natural disturbance, is another aspect of size.

Condition is an integrated measure of the composition, structure, and biotic interactions that characterize the occurrence. This includes factors such as reproduction, age structure, biological composition (e.g., presence of native versus exotic species and characteristic patch types for ecological systems), structure (e.g., canopy, understory, and ground cover in a forested community; or spatial distribution and juxtaposition of patch types or seral stages in an ecological system), and biotic interactions (e.g., levels of competition, predation, and disease).

Landscape context is an integrated measure of two factors: the dominant environmental regimes and processes that establish and maintain the target occurrence and the connectivity. Dominant environmental regimes and processes include herbivory, hydrologic and water chemistry regimes (surface and groundwater), geomorphic processes, climatic regimes (temperature and precipitation), fire regimes, and many kinds of natural disturbance. Connectivity includes such factors as species targets having access to habitats and resources needed for life-cycle completion, fragmentation of ecological communities and systems, and the ability of any target to respond to environmental change through dispersal, migration, or recolonization.

Conservation objectives that describe the intended status of each target should be articulated, i.e., characterize the desired size, condition, and landscape context that represents a viable occurrence of each target. As described below, comparison of current size, condition, and landscape context with intended status is the basis for identifying and assessing stresses—the destruction, degradation, or impairment—that afflict the priority targets, and for measuring progress towards the conservation objectives.

Stresses

It is important to understand the stresses affecting the priority conservation targets–as distinct from sources of stress–to ensure that we select the right conservation strategies. In essence, stress is the impairment or degradation of the size, condition, or landscape context of a conservation target, and results in reduced viability of the target. A source of stress is an extraneous factor, either human (e.g., activities, policies, or land uses) or biological (e.g., non-native species), that infringes upon a conservation target in a way that results in stress.

Every natural system is subjected to various disturbances. However, for our planning purposes, only the destruction, degradation, or impairment of priority conservation targets that results directly or indirectly from human causes are considered a stress. Many or most stresses are caused directly by incompatible human uses of land, water, and natural resources; sometimes, incompatible human uses indirectly cause stress by exacerbating natural phenomena. Once stresses to the priority targets are identified, they are ranked with respect to the severity and scope of damage they cause.

Sources

For each stress afflicting a given conservation target, there are one or more causes or sources of the stress. For example, nutrient loading is a stress to many aquatic ecosystems, where excess nutrients in the water draw off oxygen and, therefore, kill fish and other aquatic life. However, the nutrient loading might be caused by many different sources, such as farm fertilizers, animal feed lots, septic systems, sewage treatment facilities, or suburban runoff.

Most sources of stress are rooted in incompatible human uses of land, water, and natural resources. Such incompatible uses may be happening now (e.g., surface water diversion or incompatible livestock grazing), or may have happened in the past but left a legacy of current sources of stress (e.g., feral pigs). When multiple sources all contribute to a given stress, we want to focus our threat abatement strategies on the source or sources that are most responsible for the stress. We also want to focus on those sources that, if allowed to occur at a site, will have a long-term duration, and thereby cause long-term impacts (e.g., housing development). Once identified, sources of a stress are ranked with respect to degree of contribution to the stress and irreversibility of the stress caused by the source.

The final step in the assessment of stresses and sources is a synthesis of the individual stress and source analyses, in which the critical threats to the conservation targets are identified. A threat is actually a combination of a stress and a source of stress. For taking corrective action, the source is the thing on which threat abatement strategies are focused, under the assumption that abatement of the source will alleviate the stress and result in greater viability of the conservation target. The critical threats are those sources of stress that contribute the most to long-lasting, severe, and widespread stresses affecting multiple conservation targets.

Conservation Strategies

The conceptual framework for conservation strategies assumes that abating the critical threats will consequently alleviate the stress to the system, resulting in healthy, viable conservation targets. A threat abatement strategy focuses on abating or removing one or more sources of stress. However, in some instances, abating or removing a source of stress may not be feasible, or even if abated, the stress to the target may persist. In these situations, a restoration strategy that directly enhances or restores the viability of the target may be warranted.

Based on the identification of critical threats and the possible need for direct restoration, a set of potential threat abatement and restoration strategies should be developed. Broadly speaking, there are three complementary strategic approaches that can be deployed to abate critical threats and maintain or enhance the priority conservation targets: 1) land and water conservation, 2) public policies, and 3) compatible development alternatives.

Land and water conservation

The objective of land- and water-conservation strategies is to directly establish land and water uses and resource management that are compatible with the maintenance of the targeted systems and to ensure their short- and long-term application. This strategic approach focuses directly on resource protection and management, and includes acquisition of interests in land or water and adaptive management of public and private lands and waters.

Public policies

Some threats to biodiversity can be addressed most effectively through good public policy. For example, haphazard residential growth and urban sprawl fragment significant ecosystems across the country, not only near growing cities and suburban areas, but also in rural and coastal landscapes. To address this threat, local comprehensive plans and development standards are needed to define, design, and locate the types and amount of development that meets human community needs, protects the local environment, and generates a fair economic return. A community might provide financial incentives like tax abatements or purchase of development rights to keep land in traditional land uses, such as farming and forestry.

Because threats operate at various scales, not all threats can be addressed simply through local policies. Regional and national policy initiatives, such as the combined efforts of Maryland, Virginia, and Pennsylvania to clean up the Chesapeake Bay and revitalize its fisheries, are also needed. These policies must be founded on good information and public support.

Compatible development alternatives

Most threats to biodiversity ultimately are caused by incompatible human economic activities. To address these threats, we must often do more than appropriately use and manage resources and foster good policies that prevent incompatible activities and development. We must actively develop, promote, and implement compatible development alternatives. Compatible development is the production of goods and services, the creation and maintenance of businesses, and the pursuit of land uses that conserve biodiversity, enhance the local economy, and achieve human community goals.

Any or all of these strategic approaches may require community-based programs. Threats to the conservation targets are typically generated by incompatible human uses of natural resources and incompatible economic development. Solutions invariably require working with local landowners, community leaders, and governments. Moreover, the long-term conservation of these places will only be assured with broad-based support for conservation and compatible development within the local community. Community-based conservation programs are designed to secure short-term and long-term community support.

There is no shortage of worthwhile ideas and potential conservation strategies. Typically, however, there is a shortage of resources for implementing the strategies. Therefore, we must be hard-nosed in evaluating the benefits, feasibility, and costs of the proposed strategies in order to determine the set of strategies that can best achieve the conservation objectives within the ecological, human community, and programmatic constraints. A set of criteria for evaluating conservation strategies is presented below.

Benefits

Benefits result from abating critical threats, enhancing or maintaining the viability of conservation targets, developing conservation opportunities, and building support for conservation. Consider the marginal benefits that would arise from implementing the strategy. If the results would likely occur anyhow, without special actions by you and your conservation partners, don't rank the benefits highly. Consider 3 factors:

- Threat abatement—the degree to which the conservation strategy is likely to reduce one or more threats to the conservation targets at the site. This benefit will accrue to "threat abatement" strategies, which focus on active or ongoing sources of stress.
- Reduction of persistent stresses—the degree to which the conservation strategy is likely to reduce one or more persistent stresses (i.e., those stresses without active or current sources), and thus directly enhance the viability of the conservation target. This benefit will accrue only through "restoration" strategies, which focus on the direct reduction of stresses that have no active sources.
- Leverage—frequently, the most effective strategies are catalytic in nature, a little bit of effort or a small investment triggers positive work or resources from others and other new opportunities.

Probability of success and feasibility

All other things being equal, a program should invest in the strategies that are the most likely to succeed, in light of potentially available human and financial resources, as well as existing circumstances.

Lead person and institution

Perhaps the single most important factor of success is finding the right person to take the lead and the responsibility to implement the strategy.

Ease of implementation, and lack of complexity

Despite the best plans and the best people, there are myriad forces outside of anyone's control that can cause plans to succeed, fail, or change. The more complex the strategy, the more likely that unanticipated outside events will substantially affect the outcome. For this reason, it is wise to invest in some relatively small, simple, doable strategies. Evidence of success will then help encourage your conservation partners to undertake challenges that are more complex.

Costs of implementation

Commitment of limited discretionary resources

There are limited human and financial resources to invest in the future. Special attention should be paid to the commitment of limited discretionary resources required to implement a conservation strategy. While discretionary resources are limited, there may be opportunities to secure new resources that might be earmarked for a particular strategy.

Impact of failure

Consider the possible impact of failure, beyond the direct financial costs. For example, early failure on a high-visibility strategy might have a negative domino effect on other strategies.

Success

The conservation strategies are designed to meet two broad objectives: 1) the long-term abatement of critical threats and 2) the sustained maintenance or enhancement of the conservation targets. The target-specific objectives (i.e., desired size, condition, and landscape context) provide specificity to the second broad objective. Accordingly, our measures of conservation success assess progress towards these objectives. Specifically, the change over time in the rankings of stresses (based on severity and scope) and sources of stress (based on contribution and irreversibility), as described above, is the basis for measuring progress on threat abatement. Likewise, the change over time in the size, condition, and landscape context of the priority conservation targets is the basis for measuring progress on enhancing or maintaining the viability of priority targets.

The site planning and implementation team is responsible for designing a monitoring program that efficiently provides the appropriate information to assess size, condition, and landscape context of the conservation targets, severity and scope of the stresses, and contribution and irreversibility of the sources of stress. The team is responsible for using this monitoring information not only to assess progress towards objectives, but also to assess whether the ecological and human systems responded as expected to conservation action, to assimilate any such learning into our conceptualization and assessment of ecological systems and human context, and to reformulate conservation strategies, as warranted.

Discussion

Site conservation planning is a process that results in a specific product. The five-S approach is the framework for the process and the resulting product is a site conservation plan. The five-S framework represents a set of guiding principles for making strategic conservation decisions at sites. It can be adapted to meet the needs of local planning teams while maintaining the integrity of the guiding principles. Although developed and used by The Nature Conservancy, the process and guiding principles can be applied in any site-based conservation situation, no matter who the participants. Similarly, a site conservation plan can be designed and formatted to meet the needs and situation of the local conservation team and to best communicate with the intended audience.

Although there is not a one-to-one correspondence between the five S's and the seven components of the ERM framework, there is considerable conceptual overlap between the two frameworks. The systems, stresses, and sources components of the five-S framework parallel the issue identification and goal-setting components of ERM; the strategies component of the five-S framework is analogous to management options development and option selection; success is analogous to teaching and evaluation. These parallels reinforce the robust underlying logic common to both frameworks.

Two types of information are fundamental to the five-S process, ecological information and human context information. Information about the ecological context of the conservation targets at a site underlies the assessment of systems and stresses. Information about the human context (i.e., land use and economic factors, laws and policies, cultural attitudes, and constituencies and stakeholders) is essential for assessing sources of stress and developing effective conservation strategies.

The planning process is best deployed by an interdisciplinary team that includes the local project director who will be assuming responsibility for conserving the site, one or more scientists who are knowledgeable about the site and conservation targets, one or more participants who are knowledgeable about the local situation for conservation, and an experienced conservation practitioner who has demonstrated success at sites of similar character and complexity. When and how to involve partners, community members, and other stakeholders in the process is an important decision the planning team will have to make early in the process.

Ideally, the thought process underlying the planning should be ongoing and shared among knowledgeable participants, leading to a more thorough understanding of the five S's and the conservation requirements at a site over time. Periodically, the strategic thinking should be consolidated and the written plan updated to incorporate and document new knowledge, changing circumstances, and lessons learned.

References

Franklin JF. 1993. Preserving biodiversity: Species, ecosystems, or landscapes. *Ecol Appl* 3:202-205.

Grossman DH, Faber-Langendoen D, Weakley AS, Anderson M, Bourgeron P, Crawford R, Goodin K, Landaal S, Metzler K, Patterson K, Pyne M, Reid M, Sneddon L. 1998. International classification of ecological communities: Terrestrial vegetation of the United States. Volume I. Arlington VA: TNC.

Noss RF. 1990. Indicators for monitoring biodiversity: A hierarchical approach. *Conserv Biol* 4:355-364.

Noss RF. 1996. Ecosystems as conservation targets. *Trends Ecol Evol* 11:351.

Poiani KA, Baumgartner JV, Buttrick SC, Green SL, Hopkins E, Ivey GD, Seaton KP, Sutter RD. 1998. A scale-independent, site conservation planning framework in The Nature Conservancy. *Landscape Urban Plann* 43:143-156.

Poiani KA, Richter BD, Anderson MG, Richter HE. 2000. Biodiversity conservation at multiple scales; functional sites, landscapes and networks. *Bioscience* 50:133-146.

[TNC] The Nature Conservancy. 1997. A classification framework for freshwater communities: Proceedings of The Nature Conservancy's Aquatic Community Classification Workshop; 9-11 April 1996; New Haven, MO. Arlington VA: TNC.

[TNC] The Nature Conservancy. 2000a. Conservation by design: A framework for mission success. Arlington VA: TNC.

[TNC] The Nature Conservancy. 2000b. The five-S framework for site conservation: A practitioner's handbook for site conservation planning and measuring conservation success. Arlington VA: TNC.

Weeks WW. 1997. Beyond the Ark: Tools for an ecosystem approach to conservation. Washington DC: Island Press.

CHAPTER 14

Invited Perspective—
Protecting Life: Weaving Together
Environment, People, and Law

James R. Karr

Introduction

Environment is on everyone's mind. Concern about the effects of environmental degradation on the planet, nature, and people is widespread. At the same time, many consider environmental concerns unwarranted, even silly, but environmental issues are arguably the most important public policy controversy. Environmental concerns are now at the forefront of human affairs and failure to confront and deal with those issues is a primary global threat to human affairs.

The rapid pace of human development, driven by expanding population and advancing technology, is changing the face of Earth. The consequences of human actions now are widely understood by the vast majority of people from scholars to citizens, religious leaders to industrialists, and all walks of life and all economic sectors. As Rachel Carson (1949) noted 50 years ago, environmental problems cannot "be put off until later." They will not solve themselves "if we adopt a comfortable policy of laissez-faire."

If a proliferation of laws is any measure of reality, then environmental concerns are not being ignored. Laws are the primary mechanism available to society to defuse smoldering controversies and to establish a framework of rules within which the members of society should operate. Laws are society's response to the failure of various actors —individuals, businesses, and government entities—to perform voluntarily in ways that serve the common good. In theory, law is a thoughtful integration of social, political, and scientific knowledge designed to protect the interest of both individuals and communities.

Law evolves in response to expanding knowledge and changing societal values (Freyfogle 1998). Environmental law will continue to evolve and will likely become more effective as it grows stronger.

Risk Management: Ecological Risk-Based Decision-Making. Ralph G. Stahl Jr. et al., editors.
©2001 Society of Environmental Toxicology and Chemistry (SETAC). ISBN 1-880611-26-0

Evolution of Environmental Law

Environmental laws are passed when political leaders and citizens discover the connections among human actions, environmental degradation, and threats to human health or resources humans value. Those laws have changed rapidly in the 20th century, driven by growing awareness of environmental issues and an expanding array of challenges. Early in the century, during the administration of President Theodore Roosevelt, depletion of timber in the Great Lakes regions, contamination and excessive use of water, and degradation of western arid lands increased concerns about mismanagement of natural resources (Owen et al. 1998). But few definitive actions were taken.

By the 1930s and 1940s, grave resource problems (soil erosion, water pollution, and wildlife declines) called for serious action. Dust clouds laden with midwestern soil, for example, descended on Washington, DC, blocking the midday sun. President Franklin D. Roosevelt responded by forming the Soil Conservation Service, Civilian Conservation Corps, Tennessee Valley Authority, and other agencies (in part to halt resource degradation, but also to provide employment during a national economic crisis).

Public sensitivity and awareness of environmental challenges expanded rapidly in the 1960s, fueled by the publication of books such as Rachel Carson's *Silent Spring* (1962) and Paul Ehrlich's *The Population Bomb* (1968). The National Environmental Policy Act (1969) and the first Earth Day (1970) demonstrated that the issue engaged both Congress and the public. Laws were passed in the 1970s that expanded government responsibilities in pollution control (e.g., Clean Air, Clean Water, Toxic Substances Control, and Solid Waste Disposal Acts), conservation (Endangered Species, Marine Mammal, and National Energy Acts), and land management (Coastal Zone Management, National Forest Management, Surface Mine Control and Reclamation, and Wilderness Acts).

By the 1980s to 1990s, advances in scientific understanding and changing societal values stimulated increased efforts to enforce and strengthen these laws. Citizen monitoring programs to protect water resources became widespread, and environmental education initiatives were launched in schools from the primary grades to universities (Karr et al. 2000). Periodic efforts by Congress to turn back environmental legislation were soundly defeated. The major milestones of this period, however, were the rapid rise in nongovernmental organizations and the prominence of environmental issues to the international stage. Consider, for example, the Montreal Protocol on ozone depletion, international programs to evaluate the likely effects of global climate change, efforts to protect tropical forest and coral reefs, and international meetings such as the World Commission on the Environment and Development (1983) and the Earth Summit (1992).

Environmental law early in the 20th century sought to protect wildlife from habitat destruction, fish and game from excessive harvest, and human health from the consequences of chemical pollution. Those issues persist, and new laws and policies are evolving in response to new challenges, such as threats to political and economic security due to environmental deterioration (Myers 1993) and violent conflict stimulated by scarcity of renewable natural resources (Homer-Dixon 1999).

Despite this track record of victories for environmental protection, the environmental movement's critique of modern culture remains vague and often superficial (Freyfogle 1998). Few efforts have been made to weave the multiple goals of activist movements into a single fabric. What is the ultimate goal of environmental protection?

Sustainability has emerged as one plausible goal in the past two decades, but what is sustainability (Munasinghe and Shearer 1995)? Is the goal a sustainable economy, a sustainable development (whatever that means), or a sustainable society? With increasing frequency, protecting biological integrity or ecological health are cited as environmental goals. The idea of ecological health and protecting ecological health have become almost mainstream in science as well as public policy (Rapport et al. 1998; Pimentel et al. 2000).

Environmental Threats are Not New

History documents numerous human civilizations that developed and prospered by exploiting natural resources (Crosby 1986, Ponting 1991, Diamond 1997). Their populations grew until the resource base could no longer support them. Geographically constrained societies such as that on Easter Island fell rapidly. Constrained societies delayed the inevitable by expanding to other regions.

Through most of human history, threats to human well-being were local, temporary, and harmful mostly to the health of individuals. Medicine was developed to cure diseased or injured individuals. Over time, this focus on the individual body led to the unfortunate dissociation of human welfare from its dependence on our planet's life-support system (Karr 1993a).

By the 20th century, decisions about economic health became primary drivers of public policy. Policymakers inappropriately considered the economy as a system in which exchange values circulated in a closed loop isolated from the natural environment (Daly 1991). Economists rarely incorporated into their analyses the negative externalities associated with resource depletion or, more broadly, the depletion of the living systems that supply diverse goods and services to human economies (Daily 1997). As with medicine, the human economy was considered independent from everything outside itself (Daly 1991). This thinking, too, kept humans dissociated from their life-support systems.

People can and do live under diverse economic systems, but they live within and depend on only one life-support system. Humans are unique in several ways, but we do not live outside the laws of nature (Karr 1993b). We can no longer behave as if our dependence—and our effect—on nature is not real. First, our population size and the extent of our geographic distribution gives us an impact that is unprecedented in the history of life on Earth. Second, the evolution of culture and of technology have been instrumental in changing (not abolishing) the relationship between humans and their environment. Third, only humans are capable of recognizing the threat posed by their activities and planning to minimize or avoid those threats. Still, we are not immune from the limits of area, resources, and the environment's ability to absorb our wastes. Environmental laws are evidence that we recognize those limits. Their implementation is a test of our will.

What is Really Meant by Protecting the Environment?

Human ability to change the world outpaces the capacity of living systems to respond to these changes. As a result, the ability of Earth's systems, especially living systems, to provide the goods and services required by human society is threatened, not to mention the ability of living systems to sustain themselves. Depletion of living systems is the most important consequence of human actions. Avoiding that biotic impoverishment is the core, often unstated, goal of environmental legislation.

Biotic impoverishment is visible today in three major forms: indirect depletion of living systems through degradation of the chemical and physical environment, direct depletion of nonhuman living systems, and direct depletion of human systems (Table 14-1; Karr and Chu 1995; Chu and Karr 2001). If biotic impoverishment is the problem, then protecting the integrity of living systems must be the primary goal, or endpoint, of environmental laws. As Reynoldson et al. (1995) recently noted, "ecosystem integrity is primarily a biological concern." The challenge is defining how much change can be accommodated in a world increasingly altered by human actions. How do we reconcile the inevitable changes required to accommodate a growing human population and the proliferation of modern technology while guarding the planet from irrevocable loss of the goods and services so critical to human welfare?

Increasing recognition of these trends and their long-term consequences for human society is rapidly changing environmental policy. More importantly, these continuing ominous trends demonstrate that existing law, or its implementation, is inadequate to protect human welfare.

Table 14-1 Types of biotic impoverishment, with examples (modified from Karr and Chu 1995).

Types of biotic impoverishment	Examples
Indirect depletion of living systems	• Soil depletion and degradation (erosion, degradation of soil structure, salinization, desertification, destruction of soil biota) • Degradation of water (pollution, surface water and groundwater depletion, extinction, spread of alien taxa, homogenization of aquatic biota) • Alteration of global biogeochemical cycles • Chemical contamination (bioaccummulation, cancer, immunological deficiencies, developmental anomalies, endocrine disruption, intergenerational effects) • Global climate change and ozone depletion (global warming, alteration of rainfall distribution and amount, ozone depletion)
Direct depletion of nonhuman living systems	• Renewable-resource depletion (overfishing, excessive timber harvest) • Extinction of species • Habitat destruction and fragmentation (homogenization of biota, landscape connectivity, and mosaics destroyed) • Red tides, pest and disease outbreaks, and homogenization of crops • Introduction of alien taxa
Direct depletion of human systems	• Epidemics and emerging and reemerging diseases • Reduced human cultural diversity (genocide, loss of knowledge, loss of languages) • Reduced quality of life (malnourishment, failure to thrive) • Economic deprivation and violence • Environmental injustice (environmental racism, economic exploitation, intragenerational equity, intergenerational equity)

Laws are Not Enough

The existing framework of environmental law is weak on three counts. First, the laws are responses to narrowly conceived problems, not to integrated programs to deal with broad issues. Duplicate or competitive programs are common within and among agencies and levels of government. Each law is isolated. Pollution laws, for example, deal with individual media (soil, air, or water) rather than considering how materials move among media. Even for laws concerning the same medium, fragmentation has been the standard (e.g., legal and regulatory isolation of surface water and groundwater). Perhaps the landmark decision written by U.S. Supreme Court Justice Sandra Day O'Connor (Ransel 1995), that the separation of water quality and water quantity is an artificial distinction with no place in a law intended to give broad protection to the chemical, physical, and biological integrity of water, offers hope of change.

Second, most environmental legislation and regulations are reactive (damage control) rather than active. Problems are treated only after degradation is obvious (Karr 1990). Endangered species legislation attempts to protect species at the brink, often when it is too late. Too many people and their special-interest groups view protection of endangered species as an obstruction to progress rather than part of an overall strategy to protect the biosphere, including humans.

Third, although environmental legislation is well-intentioned and often appropriately framed, inadequate implementation prevents attainment of its goals (Karr 1990). Whether the goal is protection of soil or water, the management of national forests or protection of biodiversity, or the safeguarding of human health, laws are in place. But, implementation often falls short. Funds to protect soil and water, for example, usually go toward production enhancement (more crops, water, or fish). Programs to manage national forests subsidize timber harvest rather than protect the public's interests in standing trees and intact forests.

Programs to protect human health stopped tracking biological goals and relied instead on chemical surrogates. For some chemicals, the chosen regulatory thresholds were underprotective of people, and for other chemicals, overprotective. Regulation and cleanup of chemicals too often went beyond the level at which benefits matched costs. Bureaucracies too often used activity-based endpoints (number of permits issued, technology employed, and amount of pollution avoided) instead of evaluating actual outcomes. Rarely were ambient biological endpoints tracked in any but the most superficial ways. But if protecting biological values is the goal, biological monitoring can and should be used to track the effectiveness of regulatory decisions and incentive programs (see Karr et al. 1985; Karr 1990; Adler et al. 1993; Knopman and Smith 1993; and Reynoldson et al. 1995 for examples). Failure to evaluate the biological effects of environmental actions has often led to ecological damage as well as a squandering of public and private funds.

Protecting Ecological Resources

For years debate has raged over which ecological level (individuals, populations, communities, or ecosystems) is most effectively used as a focus to protect ecological resources. In effect, this argument is like arguing about whether to save the circulatory, digestive, or respiratory system of someone who is sick. The short answer in both cases is, we need to protect them all.

The more appropriate question is which levels provide clear, easily interpreted signals about unacceptable environmental degradation? The wise manager will seek easily interpreted signals at multiple levels. Academics might take the self-serving view that the best measure of economic health is the rate of growth of university professors' salaries. But the economy is too varied for such simplicity to guide public policy. Just as diverse economic signals provide information about the status of and trends in the economy, diverse signals should be used to track environmental quality. Examples of such signals include individual health, decline and disappearance of populations of valued or indicator species, or changes in food webs or biogeochemical cycles.

For exactly this reason, a multimetric biological index was developed, the index of biological integrity (IBI; see Davis and Simon 1995; Karr and Chu 1999; or Simon

1999 for reviews) to provide more comprehensive understanding of resource condition than is possible from examining any one biological level. Economists learned this lesson long ago, adopting multimeasure indexes such as the index of leading economic indicators or consumer price index. It is time biologists took a page from their book.

Improving Ecological Risk Assessment and Management

Risk is the probability that a specified adverse impact will occur. Historically most risk assessments focused on human health effects, usually the effects of single toxic substances from single sources. Early literature that sought to expand risk assessment to a broader biological context succeeded in expanding to other species, but still emphasized the toxicological effects of single substances. The Science Advisory Board (SAB 1990) of the U.S. Environmental Protection Agency (USEPA) and a recent Presidential/Congressional Commission on Risk (Risk Commission 1997) make a compelling case for expanding risk concepts to include ecological, as well as public health risks.

One thing is clear from these and other reports, risk assessment must be reinvented to allow and encourage managers to broaden the questions, context, and tools they apply to the nation's environmental challenges. Our first need is the need to define the specific societal goals for risk assessment, especially the risks to be averted (Karr 1995). Risk assessments that begin before defining what is at risk are likely to fail. In fact, widespread ecological degradation is a direct result of human failure to assess ecological risks properly, or even to define what is and has been at risk because of our actions. A corollary is that our future depends on our ability to recognize the deficiency and correct it. Without adequate ecological risk assessment (ERA), the prognosis for human society is grim.

Six questions are central to the advance of a truly comprehensive approach to risk assessment and risk management:
  1) What is at risk?
  2) What are the scientific and other foundations used to define what is at risk?
  3) What is the risk, and what are its consequences?
  4) How much uncertainty is associated with estimates of the risk?
  5) What data or knowledge is needed to make those risk assessments rigorous and comprehensive?
  6) What kinds of research should be done to improve risk assessment and management?

Answering these questions is crucial if we are to meet the challenges that face modern society. The threats to human-health and societal well-being no longer come primarily from the environment (Karr 1997). Rather, modern affluence has brought threats to the wider environment, ecological disruption at scales unprecedented in Earth's history, that threaten equally the well-being of human society. The depletion of the natural bounty of living systems that has nurtured and supplied modern human needs for 200,000 years is arguably the primary 21st century threat to human society. We must get a better handle on risks and uncertainties associated with that depletion.

Avoiding Unintended Consequences

Rarely did early humans need to recognize that their activities were having an effect on the supply of resources. Early technologies limited humans' ability to dominate natural systems. But, that situation has changed during the last 15,000 years with technological advances. Even our modern efforts to protect environmental health, human and ecological, have unanticipated or unintended consequences. Anyone familiar with ecological systems and natural resource management is well aware of this phenomenon. The use of pesticides to kill crop pests, alien species to control other alien species, or chlorine to kill organisms in wastewater are examples of cures that cause unanticipated illnesses. We must be more vigilant to avoid this phenomenon, an ecological analog of iatrogenesis (doctor caused) disease (Hancock 1997).

Clearly, scientists must do a better job. We must adopt a more comprehensive view of ecological risks. We must also be more effective at communicating knowledge of health challenges to citizens and policymakers. Without proper understanding of these issues, all concerned communities cannot contribute to policymaking that addresses societal well-being.

Unfortunately, one source of impediments to more-effective ERA and ecological risk management is government agencies themselves. Agencies charged with environmental protection should be protecting ecological health, but they are too often mired in conflicting missions, overlapping jurisdictions, and single-minded implementation of regulations. They count permits and compile environmental impact statements. Instead of seeking out, treating, and preventing the root causes of declining human and ecological health, agencies too often assume that keeping environmental chemicals at some predefined threshold will protect humans and other life. They are wrong.

In many circumstances, agencies responsible for environmental protection are more often environmental pollution-control agencies rather than environmental protection agencies. Proper risk assessment will never be accomplished as long as that mindset dominates the agencies charged with protecting the public's interest in ecological health.

Whatever the framework for assessing ecological risks, each step must be informed by data from biological monitoring (Karr and Chu 1997, 1999). For accurate, relevant ERA, the measurement endpoints (what is measured) and the assessment endpoints (the ecological goods and services society seeks to protect) must be explicitly biological. Biological monitoring provides more useful information about actual environmental quality than chemical and physical measures alone (Keeler and McLemore 1996) because biological attributes are one step closer to the factors that constitute environmental quality for living things. Microeconomic models based on chemical levels as surrogates of environmental quality may be useful for approximating the costs of pollution control, for example, but they are limited in their ability to explain the ecological, explicitly biological, damage caused by that pollution (Keeler and McLemore 1996). Economic models incorporating biological measures, on the other hand, can contribute more accurately to a whole-system approach to risk management.

Ecological risk assessment has grown in complexity and stature with recognition that direct risks to human health are not the only environmental risks that society faces. For ERA to fulfill its potential, more careful thought must go into defining the specific societal goals for ERA, especially the risks to be averted. As we enter a new millennium, we can no longer see ecological risks so narrowly. We know that direct threats to human health from chemical contamination are only a small segment of the problem. We know that our failure to provide public policy programs to treat other risks has mortgaged the human future just as it has wasted resources on narrow regulatory programs.

The primary goal of ERA should be the protection of the fabric of life. We can protect that fabric by recognizing and making decisions based on the reality of connections between humans and their environment. The implementation of sound environmental statutes is vital to that protection.

Acknowledgments—Partial support for this work came from the Consortium for Risk Evaluation with Stakeholder Participation (CRESP), established under U.S. Department of Energy Cooperative Agreement #DE-FC01-95EW55084.S. Dr. Elen W. Chu provided a critical review of this paper.

References

Adler RW, Landman JC, Cameron DM. 1993. The Clean Water Act 20 years later. Washington DC: Island Press.

Carson R. 1949. Lost worlds: The challenge of the islands. In: Lear L, editor. 1998. Lost woods: The discovered writing of Rachel Carson. Boston MA: Beacon Press.

Carson R. 1962. Silent spring. Boston MA: Houghton-Mifflin.

Chu EW, Karr JR. 2001. Environmental impact, concept and measurement of. In: Levin S, editor. Encyclopedia of biodiversity. Volume 2. Orlando FL: Academic Press. p 557-577.

Crosby AW. 1986. Ecological imperialism: The biological expansion of Europe: 900-1900. New York NY: Cambridge University Press.

Daily GC, editor. 1997. Nature's services: Societal dependence on natural ecosystems. Washington DC: Island Press.

Daly HE. 1991. Elements of environmental macroeconomics. In: Costanza R, editor. Ecological economics: The science and management of sustainability. New York NY: Columbia University Press. p 32-46.

Davis WS, Simon TP, editors. 1995. Biological assessment and criteria: Tools for water resource planning and decision making. Boca Raton FL: Lewis Publisher.

Diamond JM. 1997. Guns, germs, and steel: The fates of human societies. New York NY: Norton.

Freyfogle ET. 1998. Bounded people, boundless lands: Envisioning a new land ethic. Washington DC: Island Press.

Hancock T. 1997. Ecosystem health, ecological iatrogenesis, and sustainable human development. *Ecosyst Health* 3:229-234.

Homer-Dixon TF. 1999. Environment, scarcity, and violence. Princeton NJ: Princeton University Press.

Karr JR. 1990. Biological integrity and the goal of environmental legislation: Lessons for conservation biology. *Conserv Biol* 4:244-250.

Karr JR. 1993a. Protecting ecological integrity: An urgent societal goal. *Yale J Int Law* 18(1):297-306.

Karr JR. 1993b. Using Biological criteria to ensure a sustainable society. In: Watershed resources: balancing environmental, social, political and economic factors in large basins. Conference Proceedings Oregon State University Department of Forest Engineering; 14-16 October 1992; Portland, Oregon. Corvallis OR: Oregon State University. p 61-70.

Karr JR. 1995. Risk assessment: We need more than an ecological veneer. *Hum Ecol Risk Assess* 1:436-442.

Karr JR. 1997. Bridging the gap between human and ecological health. *Ecosyst Health* 3:197-199.

Karr JR, Allan JD, Benke AC. 2000. River conservation in the United States and Canada. In Boon PJ, Davies BR, Petts GE, editors. Global perspectives on river conservation. London, UK: John Wiley and Sons, Inc. p 3-39.

Karr JR and Chu EW. 1995. Ecological integrity: Reclaiming lost connections. In: Westra L, Lemons J, editors. Perspectives on ecological integrity. Dordrecht Netherlands: Kluwer. p 34-48.

Karr JR and Chu EW. 1997. Biological monitoring: Essential foundation for ecological risk assessment. *Hum Ecol Risk Assess* 3:993-1004.

Karr JR and Chu EW. 1999. Restoring life in running waters: Better biological monitoring. Washington DC: Island Press.

Karr JR, Heidinger RC, Helmer EH. 1985. Sensitivity of the index of biotic integrity to changes in chlorine and ammonia levels from wastewater treatment facilities. *J Water Pollut Control Fed* 57:912–915.

Keeler AG, McLemore D. 1996. The value of incorporating bioindicators in economic approaches to water pollution control. *Ecol Econ* 19:237–245.

Knopman DS, Smith RA. 1993. Twenty years of the Clean Water Act. *Environment* 35(1):16–20, 34-41.

Munasinghe M, Shearer W, editors. 1995. Defining and measuring sustainability: The biogeophysical foundations. Washington DC: The World Bank.

Myers N. 1993. Ultimate security: The environmental basis of political instability. New York NY: Norton.

Owen OS, Chiras DG, Reganold JP. 1998. Natural resource conservation. 7th Edition. Upper Saddle River NJ: Prentice-Hall.

Pimentel D, Westra L, Noss RF. 200. Ecological integrity: Integrating environment, conservation, and health. Washington DC: Island Press.

Ponting C. 1991. A green history of the world: The environment and the collapse of great civilizations. New York NY: St. Martin's Press.

Ransel KP. 1995. The sleeping giant awakes: PUD No. 1 of Jefferson County v. Washington Department of Ecology. *Environ Law* 25:255-283.

Rapport D, Costanza R, Epstein PR, Gavdet C, Lewis R. 1998. Ecosystem health. Malden MA: Blackwell Science.

Reynoldson TB, Bailey RC, Day KE, Norris RH. 1995. Biological guidelines for freshwater sediment based on benthic assessment of sediment (the BEAST) using a multivariate approach for predicting biological state. *Aust J Ecol* 20:198-219.

Risk Commission (Presidential/Congressional Commission on Risk Assessment and Risk Management). 1997. Framework for environmental health risk management. Washington DC: Presidential/Congressional Commission on Risk Assessment and Risk Management.

[SAB] Science Advisory Board. 1990. Reducing risk: Setting priorities and strategies for environmental protection. Washington DC: USEPA. SAB-EC-90-021.

Simon TP, editor. 1999. Assessing the sustainability and biological integrity of water resources using fish communities. Boca Raton FL: CRC Press.

CHAPTER 15

Summary, Conclusions, and Recommendations

Ralph G. Stahl Jr., Anne L. Barton, Robert A. Bachman, James R. Clark, Peter L. de Fur, Jessica Glicken Turnley, Stephen J. Ells, Charles A. Pittinger, Michael W. Slimak, Randall S. Wentsel

Summary and Conclusions

The Society of Environmental Toxicology and Chemistrys (SETAC) Framework for Ecological Risk Management Technical Workshop held 23-25 June 1997 in Willamsburg, VA culminated nearly 2 years of prior discussion and interaction among risk assessors and risk managers in the regulated and regulatory communities. This book represents another 2 years of further discussion on and compilation of the workshop's products (see also Pittinger et al. 1998), in addition to gaining written perspectives and examples from those not participating in the original efforts. As we've learned, there was clear consensus at the workshop that the topic of ecological risk management (ERM) was timely, if not overdue, making the publication of this book another important step in advancing the science and policy of ecological risk assessment (ERA) and ERM. The major product of the workshop has been a framework for ERM (Chapter 2), outlining a participative, decision-making process stressing input from multiple stakeholders. The ERM framework closely complements the U.S. Environmental Protection Agency's (USEPA's) (1992) model for ERA by offering guidance on the substantive nature of interactions between risk assessors and risk managers, both preceding and following an ERA (Chapters 1, 3, 4, 5, 6).

Workshop participants felt strongly that ERM concerns representing the diversity of stakeholder inputs needed to be considered early in the design of an ERA to ensure the assessment optimally supports sound decision-making. In addition to the risk assessment itself, workshop participants identified a number of other key considerations for ERM decisions (Chapters 3, 4, 5). It was recognized that sound and acceptable ERM decisions are supported by a process which effectively integrates social, political, economic, and technical interests and concerns. In this regard, similarities to human-health risk management were apparent, despite the greater complexity of ecological risks resulting from species diversity, scales of biological organization, and numbers of endpoints (Chapters 1 and 4).

It was the consensus of the workshop participants that narrow and prescriptive criteria for acceptable ecological risk should not be offered, as decision criteria and the acceptability of management decisions were seen as highly contextual and case-specific. However, workshop participants articulated a number of broadly-applicable recommendations based upon their practical experience. They viewed sound decisions as those which are meaningful and relevant to the issue at hand and avoid causing equal or worse problems in other media or contexts.

Innovation and flexibility in decision-making were seen as highly valued traits of decision-making. The underlying role of the legal process in environmental management, particularly in the U.S., was viewed as a barrier to both scientific and social discourse. The fear of litigation and its associated stigma tend to stifle creativity in some cases, and once the legal process begins, the interface between litigants can become an impediment to science-based exchanges. It was generally agreed that benefits of risk reduction or risk avoidance need to justify costs to society, though valuation of natural resources and ecological risks was seen as a highly subjective and inconsistent process.

This SETAC Technical Workshop Framework for Ecological Risk Management was rather unique in the broad diversity of affiliations and fields of expertise represented among the participants. The steering team endeavored to bring together a healthy contingent of natural resource managers from federal and state wildlife conservation programs, not the typical SETAC participants but whose expertise and knowledge were an added benefit to the workshop. An a priori hypothesis by the steering team that resource management goals and criteria are consistent with those of conventional risk scientists was indeed validated, as the resource managers at the workshop contributed an immensely valuable and practical perspective to the discussions on decision-making criteria.

Recommendations

As ERA gains wider use in the governmental (USEPA 1992, 1998) and business communities (CMA 1997; AIHC 1998), so emerges a parallel need for fundamental guidance in ERM.

Recommendation 1:

Begin a multi-stakeholder process for developing ecological risk management guidance for the U.S.

Good guidance would help to ensure consistent, science-based management decisions, but not be a rigid menu of criteria or a limited set of options. Moreover, it would provide a flexible process to help decision-makers set protection goals and to decide on a case-by-case basis what actions to take. Today, there is no simple guide for ecological risk managers to account for changing social values, regional perspectives, or the diversity of issues associated with ERM. Transparent and coherent ERM criteria can offer benefits for regulators and the regulated community alike by clarifying the diversity of management concerns, e.g., environmental, economic, legal and social, and the related decision criteria for each. Ideally, ERM guidelines will define how to build a consensus process through which risk managers from the public and private sectors could operate to identify 1) the ecological resources to protect, 2) the circumstances under which they need protection, 3) the degree of protection, and 4) the means to achieve that protection. At a minimum, these could provide guidance on decisions for new chemicals, waste sites, natural resources, and fill the void in existing guidance. Such guidance should have the benefit of broad stakeholder input and be capable of providing flexible and prudent approaches to ERM.

Several recent initiatives have helped underscore and advance the need for ERM guidelines. In 1994, the USEPA (USEPA 1994) published a thought-provoking essay that illustrated the diversity of approaches that the USEPA regions and offices have applied to ERM. Later, the USEPA (USEPA 1995) published an internal guide for managers faced with making decisions on ecological risks. More recently, the USEPA (USEPA 1997) published management approaches and listed types of ecological entities to be considered. Likewise, the USEPA (Luftig 1998) published draft principles for ERM at Superfund sites. All of these can be good starting points for ERM guidance development.

On 19 January 1999, the USEPA held a Colloquium on Setting Ecological Risk Management Objectives in Washington, D.C. Participants at the colloquium largely agreed on a recommendation to the USEPA that ERM guidance, developed with participation by all stakeholders, would be a timely and essential complement to the agency's guidance on ERA. It was further recognized that the work products of SETAC's Williamsburg Workshop, together with the USEPA's documents discussed above, provide a strong foundation in ERM on which to build.

While there is a lot to be gained by having guidance, potential pitfalls exist. What the guidance should not do is establish a rigid procedure for ERM. Nor should it attempt to dictate a fixed balance of environmental, economic or social considerations because these must be measured in the context and value systems in which they occur. It should allow for stakeholder input, but not delay decision-making by introducing unnecessary layers of authority, excessive steps for approval, or become an excuse for inaction.

Recommendation 2:

Begin a national dialog on approaches to valuing ecological resources and the levels at which these resources should be protected

Even though the concerns and values encompassing ecological risks are broad, there is little that is fundamentally different in the basic approaches used to assess risks to humans versus the environment. However, the variety, breadth, and scale of ecological systems are inherently more vast, which complicates both ERA and ERM.

Throughout human history, society has developed and implemented processes and practices for measuring and managing risks to humans. The value of ecological resources and how best to manage risks to them has proven much more difficult to understand. This issue is the crux of the discussion that we have yet to undertake nationally and should be an important aspect of ERM guidance deliberations among the governmental, environmental, and business communities.

The federal mandate for managing and protecting the nation's biotic and abiotic resources is embodied in various pieces of environmental legislation. Sweeping statements in the legislative preambles are inspirational, but provide little practical guidance on exactly what fishable and swimmable actually means or how to achieve it. Because the key environmental statutes were for the most part crafted over a 20-year period (i.e., the 1970s and 1980s) by different congresses under different administrations, it is not surprising that environmental-management goals and standards (when they are articulated) vary widely across media, stressor types, and scales of biological organization (e.g., species, ecological communities, or ecosystems).

As noted above, legislation by itself rarely provides adequate practical guidance on how good management can be accomplished (the Endangered Species Act being a notable exception) (Endangered Species Act of 1973; Public Law 93-205). This may not be problematic however because prescriptive, statutory guidance can hamstring regulators and the regulated community alike. Added to this is the need of a democratic society to debate what constitutes adequate and reasonable protection, and when, where and how to administer sound management. Ultimately, this results in a hard-to-achieve balance among shifting societal norms and expectations which may appear to be an insurmountable obstacle to developing ERM guidance for the U.S. Nevertheless, the goal of ERM guidance should be to provide a framework and process where these issues can be addressed and appropriate protection afforded.

Recommendation 3:

Design and conduct a SETAC-style workshop to further refine the elements of the proposed ecological risk management framework and test them using case studies from new chemicals, waste sites and natural resource evaluations.

To determine the utility of the proposed ERM framework, it is important to test it with a diverse array of case studies from the three main application areas—new chemicals, waste sites, and natural resource evaluations. This step could best be accomplished through a SETAC-style Pellston Workshop that would bring together scientists and managers from the academic, business, governmental, and environmental communities. The product of this workshop would be a further refinement of the proposed ERM framework and, if needed, changes to it that would account for issues that arose during the review. This effort might also be best done before national guidance is attempted as it may illuminate additional issues that need to be resolved.

References

[AIHC] American Industrial Health Council. 1998. Ecological risk assessment: Sound science makes good business sense. Washington DC: American Industrial Health Council. 14 p.

[CMA] Chemical Manufacturers Association. 1997. Ecological risk assessment: A tool for decision making. Arlington VA: CMA. 20 p.

Luftig S. 1998. Draft ecological risk management principles for Superfund sites. OSWER Directive 9285. p 7-28.

Pittinger CA, Bachman R, Barton A, Clark JR, deFur PL, Ells SJ, Slimak MW, Stahl RG, Wentsel RS. 1998. A multi-stakeholder framework for ecological risk management: Summary from a SETAC Technical Workshop. *Environ Toxicol Chem* 18(Supplement).

Presidential/Congressional Commission on Risk Assessment and Risk Management. 1997. Final Report. Volume 1. Washington DC. 64 p.

[USEPA] U.S. Environmental Protection Agency. 1992. Framework for ecological risk assessment. Washington DC: USEPA. EPA-630-R-92-001.

[USEPA] U.S. Environmental Protection Agency. 1994. Managing ecological risks at EPA. Issues and recommendations for progress. Washington DC: USEPA. EPA-600-R-94-183.

[USEPA] U.S. Environmental Protection Agency. 1995. Ecological risk: A primer for risk managers. Washington DC: USEPA. EPA-734-R-95-001.

[USEPA] U.S. Environmental Protection Agency. 1997. Priorities for ecological protection: An initial list and discussion document for EPA. Washington DC: USEPA. EPA-600-S-97-002.

[USEPA] U.S. Environmental Protection Agency. 1998. Guidelines for ecological risk assessment. Washington DC: USEPA. EPA-630-R-95-001F.

Abbreviations

ASTM	American Society for Testing and Materials
ARAR	Applicable, relevant, and appropriate requirement
AHM	Adaptive harvest management
CERCLA	Comprehensive Environmental Response, Compensation, and Liability Act
CWA	Clean Water Act
CC	Concern concentration
dSAYs	Discounted service-acre-years
ERA	Ecological risk assessment
ERM	Ecological risk management
ESA	Endangered Species Act
EO	Executive Order
EPW	Evaluation for planned wetlands
EDC	Ethylene dichloride
EEC	Estimated environmental concentration
FIFRA	Federal Insecticide, Fungicide and Rodenticide Act
GIS	Geographic information system
HHRA	Human health-risk assessment
HEA	Habitat Equivalency Analysis
NRC	National Research Council
NOEC	No-observed-effects concentration
NCP	National Contingency Plan

Ecological Risk Management: A Framework for and Approaches to Ecological Risk-based Decision-making. Ralph G. Stahl, Jr. et al., editors. ©2000 Society of Environmental Toxicology and Chemistry (SETAC). ISBN 1-880611-26-0

NRDA	Natural resource damage assessment
OPPT	Office of Pollution Prevention and Toxics
OPP	Office of Pesticide Programs
PMN	Premanufacture notification
ROD	Record of Decision
SETAC	Society of Environmental Toxicology and Chemistry
SAR	Structure activity relationship
SRC	Migratory Bird Regulations Committee
TSCA	Toxic Substance Control Act
USEPA	U.S. Environmental Protection Agency
USFWS	U.S. Fish and Wildlife Service
WIPP	Waste Isolation Pilot Plant

INDEX

A

Adaptive management, 120–128, 168
Administrative Procedure Act, 115
Aix sponsa, *113–123, 127–128, 131*
American Chemistry Council, Responsible Care
    Program, 9, 139, 147
American Society of Testing and Materials
    (ASTM), 2–3, 14, 23
Analysis of ecological risk assessment, 16, 17–18
Anas *spp.*, *113–131*
Applicable, relevant, and appropriate require-
    ments (ARARs), 107
Aquatic Dialogue Group, 13
Aquatic Nuisance Species Task Force, 133
*ASTM. See American Society of Testing and
    Materials*

B

Barium, DuPont Superfund site, 149
Biodiversity, 79, 163, 169
Biological monitoring
    importance, 182–183
    of waterfowl, 118–120, 128
Biological stressors. See Invasive species
Biotic impoverishment, 178, 179
Birds, adaptive regulation of hunting, 113–128,
    131
Black carp, 133–136
Branta canadensis, *113–123, 127–128, 131*

C

Calcasieu estuary restoration, 157–161
Canada geese, 113–123, 127–128, 131
Case studies, to test ERM framework, 190–191
*CERCLA. See Comprehensive Environmental
    Response, Compensation and Liability Act*
Chemicals. See Pesticides; Stressors
Chesapeake Bay Program, 63, 169
Civilian Conservation Corps, 176
Clean Air Act, 96
Clean Water Act, 8, 95, 96
Colloquium on Setting Ecological Risk Manage-
    ment Objectives, 189
Common reed, 151

Communication
    corporate, 138, 151
    definition, 68
    with the public, 6, 12
    between risk assessor and risk
        manager, 6, 22–23, 26, 108,
        109, 136
    skills of a risk manager, 12, 28, 37
    with stakeholders, 37–38, 66–68
Community development, 169–170
Community-level risk management, 10, 44, 49,
    50, 165
Comprehensive Environmental Response,
    Compensation and Liability Act (CERCLA),
    8, 11, 42, 61, 105–108
Compromise, 30. See also Tradeoffs
Conservation strategies, 168–171
Consumer Product Safety Act, 96
Corporate aspects
    employee commitment, 139
    multiple connotations of "safe," 144–
        146
    role of ERM, 9, 10, 137–148
    role of stockholders, 10
Costs and benefits. See Economic aspects
Cultural aspects
    relationship between humans and their
        environment, 177–178
    of risk management, 1, 2, 22
    "safety" of a technology, 145–146

D

Data acquisition, verification, and monitoring,
    in ERA, 22, 63–64
Data compilation and analysis
    Calcasieu estuary restoration, 158–161
    corporate context, 142–144
    DuPont Superfund site, 151–153
    in ERM framework, 27–28
    interspecies extrapolation, 27, 93
Data gaps, 8, 27, 53, 54. See also Uncertainty
Decision criteria, 26–27, 53–54
    corporate, 144–146
    in pesticide premanufacture notices, 99
Decision goals, 4
Decision implementation. See Implementation
Decision-making
    Calcasieu estuary restoration, 161
    under CERCLA, 105–108
    corporate, 9, 10, 137–148

U

V

W